How to Build a

With Plans

The Science of Living with Less Series
Mendon Cottage Books

JD-Biz Publishing

Our books are available at

1. Amazon.com
2. Barnes and Noble
3. Itunes
4. Kobo
5. Smashwords
6. Google Play Books

Table of Contents

Chapter One: Introduction to Tiny Houses

The Tiny House Movement

The tiny house movement is gaining momentum and popularity for it cost effectiveness and simplicity. People who support this movement are happy to live in a downsized home -- their tiny house is a lifestyle that they adopt/adapt and are happy in doing so. While one might argue the disadvantages of living in a tiny space, the financial freedom that it brings along cannot be disregarded.

The average size of a house in the United States of America was 1,500 square feet in the 1950's; now, when the family sizes are shrinking, the average house size is 2,400 square feet. The lavish spaces in a 2,400 square feet house lies dormant for most part of the year. One such classic example is the living room, which serves it purpose for a couple of hours during the Christmas season... rest of the time its existence is hardly noticed or acknowledged; ditto for the multiple bathrooms, the huge dining room, bedrooms, and garage parking. The cost one pays for a huge home that is hardly used is hefty -- a mortgage that stretches almost until the end of one's retirement. Add to this the maintenance cost and the carbon footprint of a huge house!

Chapter Two: Building Codes and Zoning Requirements

Building Codes

Before proceeding with the purchase of a trailer and constructing your tiny house -- THOW (Tiny House On Wheels), it is important to understand the building and zoning codes set forth by your jurisdiction. Trailer houses are built on platform foundations as it helps in mobility; these homes are built according to codes -- the only exemption being that one doesn't need a building permit to build a house on a trailer platform.

The local city hall has a planning department, which helps people understand the coding and zoning restrictions in your locality. Before paying them a visit prepare a set of questions and clear all your doubts. Remember -- when it comes to rules and regulations, any doubt (however trivial it might be) is worth getting cleared. Since you are planning to build a THOW, be prepared to expect the unexpected! People working in the planning department may not be able to answer all your questions and clear your doubts, but they will direct you to the DMV department.

With a THOW, you will face problems with parking and not with structural design of your house. You must register your house under one of the following categories:

1. Semi trailer
2. RV (Recreational Vehicle)
3. Mobile home

Zoning Codes

You will face legal issues with parking your THOW. Even though some counties have changed their local zoning codes that allow people to live in their RV full time, it is a long way before proper codes for THOW are implemented into Building and Zoning Codes. Here are a few things to check with your local zoning authorities:

A. Can you park your THOW (RV) on your property without a main house?

B. Are you allowed to live in your RV permanently

There is absolutely no point fighting with the authorities about the codes and zoning regulations. If you hit a rough weather trying to find a parking for your THOW, here are a few pointers to consider:

1. Try locating a backyard where sheds and/or THOW are allowed
2. Post a WANTED advert in a newspaper
3. Try Craigslist WANTED as well
4. Try RV parks
5. If you are active on the social media -- post a query or the WANTED ad on Facebook/Twitter; you can also create a group or join an already existing group for THOW.

Chapter Three: Buying and Registering a Trailer

Buying a Trailer

Once you have decided to build a THOW, an important decision to make is about purchasing a trailer. You can buy either a new trailer or purchase a second hand trailer. The latter will save you a considerable amount of money -- you can invest this little savings to buy quality products for your home.

Most tiny homes are built on Flatbed trailers; they are also called lowboy trailers or utility trailers. Your concern here must be about the type of hitch used on the trailers. There are two types of hitches, each with their own set of advantages and disadvantages:

A. Bumper Pull:

This is the most common type of of hitch, also called drag or tagalong trailer. The advantages of a bumper pull trailer are:

- Its compact size
- Costs less than a gooseneck trailer
- It can be towed using a SUV, CUV, or a motorhome
- The combined weight of a bumper pull trailer and the vehicle will not exceed 10,000 pounds. (Please check with your state's DMV for the limits)

Disadvantages:

- Due to its compact size, the living space is reduced
- You can't haul heavier loads (please ensure that your trailer and truck are rated properly for the loads that you carry)

B. Gooseneck Trailer:

Advantages:

- The sway is considerably reduced since the weight of the trailer is over the truck's rear axle; this leads to better stability
- It is easier to manoeuvre tighter spaces, as the turn radius is tighter

Disadvantages:

- It is heavy enough to be classified as commercial vehicle
- It can be hauled only with a pickup truck or something similar
- It requires a special hitching system installed in the bed of the pickup truck
- A gooseneck trailer ball must be installed in the truck beds

Chapter Four: Tools and Materials required

Tools and Materials

After having chosen your trailer, the next step is to build your tiny house on the trailer. However, before proceeding to the construction part, you must get the tools ready. People work differently -- some require all the tools and equipments available in the store... to build a small box! There are others who manage to build a house with bare minimum tools.

Here is a list of tools that are necessary during the construction process:

1. Skill saw, table saw, and jigsaw
2. Files
3. Plyers
4. Hammer
5. Wrench
6. Measuring tape
7. Marking pencils and colors
8. Spirit level
9. Gloves, goggles, respirators, sound protectors, good boots, and first aid kit
10. Box cutter
11. Chisel
12. Drill and drill bits

Chapter Five: Foundation

Details

As you are building a mobile house, you will not need a foundation -- your trailer's flat bed functions as the foundation; you can remove the extra decking to leave 24" gap between the remaining boards. This reduces the weight. After the extra decking is removed, the next step is to protect your house from moisture, rain water, and rodents. For this purpose, you need to add a layer of flashing.

To do this:

1. Use aluminium flashing
2. Use a radiant barrier such as EcoFoil
3. Use a galvanised metal flashing

A vapor barrier too can be added on top of the flashing.

Alternatively, instead of flashing the trailer, you can use roof felt to cover the deck without removing the extra pieces; roof felt will keep your house safe from water and bugs. If your trailer has a porch, do not add anything under the porch.

Chapter Six: Flooring

Framing

Once your trailer is ready with the flashing, the next step is framing. For framing your THOW, you must follow the plans (available at the end of this book).

There are few basic things to keep in mind before you start framing. These are general tips that will help in framing any kind of house:

1. All joists have a crown; align the joists so that the crowns are in the same direction

2. Mark the crown as you are aligning them; this way you can avoid mistakes later

3. When you nail the joists into place, the crown must face up

4. If the joists overlap an interior wall, cut the joists approximately 2" beyond the wall

5. Overlapping joists must not protrude more than 6" beyond the wall

Subfloor framing:

For framing the subfloor, you can opt for 2 methods:

A. Directly frame the lumber onto the flashing

B. Prepare the framing as three different units and assemble it on the flashing

The advantage of the 2nd method is that it is easier to frame smaller units and attach them to the flashing. It is also easier to undo the framing, if need be. For this method, you have to prepare 3 frames - 1 each for either ends and 1 for the center.

Usually 2x4 lumber is used for framing the floors. You can either use screws or nails for framing. Screws will not pull out but can snap whereas nails do the opposite i.e., they don't snap but can pull out. Thus, it is a good idea to use a mix of screws and nails.

When you frame the floor, the joists must be at 4' intervals - this is where the sheathing will meet.

Once the floor framing is ready, it is time to add insulation in the gaps. Following are the types of insulation to consider:

- ✓ Fibreglass - most tiny house builders opt for fibreglass as insulation material. It is fairly inexpensive and easy to install. However, you must purchase fibreglass that is considerably thick to provide good insulation.
- ✓ Spray foam - you can use polyurethane spray foam, as it provides excellent insulation. It is made of high density closed-cell foam and requires certain level of expertise to install it. It is advisable to hire professional help in installing the spray foam.
- ✓ Styrofoam boards supplemented with spray foam - it is slightly more expensive than fibreglass but can be installed without professional help.
- ✓ EPS or Extruded Polystyrene - this is the most popular choice of insulation for THOW. You must use a 3" rigid foam board before installing the subfloor sheathing.
- ✓ Other insulation materials to consider are sheep wool, stone wool, and cellulose - sheep wool though organic will be sprayed with fire retardant and insect repellents; stone wool is made from diabase volcanic rock melted with coke and slag in a furnace. Cellulose (wet spray) is made from newspapers and requires a high power blower for proper installation - this insulation is not recommended for tiny houses due to the difficultly in installing the insulation.

Before proceeding to laying of the subfloor, install rough wiring and plumbing according to the plans.

Subfloor

After adding your preferred insulation to the frame, the next step is to lay the sub flooring - it stabilises your flooring. For a THOW, the floor carries the majority of the weight and provides structural integrity. There are three types of sub flooring to choose from:

1. OSB or Oriented Stran Board
2. Plank
3. Plywood

1. OSB or Oriented Stran Board: This is a cost effective sub flooring and the simplest of the three. OSB is compressed wooden chips that are glued together to form 4'x8' sheets. You must either glue or nail the OSB sheets to the floor joists.

2. Plank: A plank sub flooring consists of yellow pine boards; these are usually 3/4"x4"-8" wide. You can install the planks by nailing them directly to the floor joists. A plank sub floor will loosen with time due to the expansion and contraction of wood. Thus, it is necessary to use 2.5" or 3" deck screw.

3. Plywood: These are usually made of a pine veneer, which is laminated and glued together to form 4'x8' sheets. You can opt for waterproof plywood sheets for the entire house or use the regular sheets. It is advisable to use waterproof sheets in areas of water usage (kitchen, bathroom).

Sheathing

Here are a couple of tips for sheathing the subfloor (or any part of your house):

1. Buy sheathing material only when you are ready to sheath

2. Buy good quality material; buying a cheaper material will save you money today but will cause more problems in the future

When you frame the floor, the joists must be at 4' interval - this is where the sheathing will meet.

Steps to sheath a subfloor:

1. Draw a line with chalk at 4'-1/4" on joists

2. Apply glue to the joists (if mentioned in the plan)

3. If you apply glue, you must nail the sheathing before the glue dries

4. Centre the sheathing at the last joist; the free end of the sheathing must align with the chalk line

5. Nail both the corners of the sheathing at the last joist

6. Continue to nail the sheathing aligning with the joists and the chalk line

7. When you sheath the remaining rows, set to existing sheets and allow a 1/8" gaps

Chapter Seven: Walls

Framing

With the subfloor sheathing done, you can now concentrate on framing the walls and roofs. Before you start framing the walls, it is advisable to have your door and window frames ready -- this allows you to test fit the frames.

If you have multiple people to help you with your construction, entrust the task of cutting the lumber to one person. This results in efficiency and also ensures consistent cuts. If you are working alone, cut all the required pieces of lumber before you start framing. Usually 2x4 lumber is used for wall framing.

It is easier to frame and sheath the walls in horizontal position and later lift it to a vertical position.

Tips for wall framing:

1. Use a chalk and mark the wall framing layout on the floor; this will reduce errors
2. Mark the inside edge of the wall at the corners
3. Now mark the interior wall locations this will be your bathroom, windows, and door opening
4. Measure and cut the top and bottom plates of the wall
5. Spread the studs
6. Nail the header to studs
7. Next nail the top plate to studs and headers
8. Nail the bottom plate to studs

9. Finally nail the double plate to the studs - using a double plate makes your construction sturdier

Squaring the wall:

Once the framing is done, the next step is to square the wall. For this, secure the bottom plate of the wall and move the top plate until the diagonal dimensions of the wall are equal. Secure the wall into the floor by driving two nails through the double plate. Now nail the inside of the bottom plate.

At this point you can choose to sheath the wall framing before you stand up the wall; alternatively you can sheath the walls after standing up the wall. Plumb and line the wall once it is upright.

Chapter Eight: Roofs

Framing

Framing a roof is the most difficult part of any building construction and involves calculations to determine the length of the rafter. Following are some terminologies related to the rafter that will help you in framing:

1. Span - this is the distance between two load bearing walls; it is measured from the outside of the walls.
2. Run - horizontal distance from the outside wall to the center of the ridge
3. Rise - vertical distance for which the roof rafter board extends upwards
4. Slope - ratio of rise to run; it is always expressed as inches per foot
5. Pitch - incline of the roof, which is obtained by dividing the rise by the span; it is expressed as a fraction
6. Pitch angle - vertical angle that represents the pitch of the roof
7. Framing point - the point where the centre line of connecting rafters, ridges, hips, or valleys meet
8. Common rafter - a rafter that runs from a wall straight to a ridge board
9. Jack rafter - rafter that runs to a hip rafter or a valley rafter
10. Hip rafter - the rafter that joins jack rafters on the roof; hip rafters run in between the jack rafters
11. Valley rafter - rafter at the inside corner of a roof; it runs between and joins the jack rafters from each side
12. Connection angle - horizontal angle at the end of a rafter; this angle is used for connecting with other rafters, hips, valleys, or ridge boards

13. Fascia - a board nailed at the end of the rafters at the eaves
14. Overhang - extension of roof beyond the wall

P.S. Pitch and slope are used interchangeably this is incorrect; they mean different things.

Types of Roofs for Tiny Homes:

Following are the roof types that can be used in the construction of a tiny house:

A. Gable roof - this is the easiest roof to construct with a simple design. A Gable roof has a ridge in the center and slopes on either side. While it is easy to construct and economical, it also wastes maximum space. This roof has to be reinforced for strength but doing so will take away headroom in your loft (if you are planning to have one)

B. Hip roof - this is a strong roof, as it has 4 hip rafters but it is very difficult to construct a hip roof. The roof has 4 sloping sides and the hip rafters run at an angle of 45 degrees from each corner to the ridge

C. Shed - this roof has only one slope and is the best option for house with a loft. With this type of roof, you can add extra windows at the peak side of the roof and also harvest rainwater

D. Saltbox roof - this is a two sided roof with the advantage of loft space; it is similar to a shed but has an off-center peak. Since the load is not centred in this roof, it is structurally weaker than the other roof types

E. Arched roof - this is the strongest of all roof types but very difficult and expensive to construct; go for an arched roof if you want a long lasting roof and have the money

F. Flat roof - try to avoid this type of roof, as it is prone to damage from debris, and water

Determining the Rafter Length

The length of the rafter is dependent on the slope and the span and following are the ways to determine the rafter length:

1) Pythagoras theorem
2) Framing Square rafter method
3) Framing Square stepping method
4) Computer software method
5) Diagonal percent method
6) Chalking line duplication method

Using Pythagoras Theorem:

This theorem states that in a right angled triangle, the square of the hypotenuse is equal to the sum of the square of the other two sides. If A & B are sides of a right angle triangle and C is the hypotenuse, we have:

$A^2 + B^2 = C^2$

When the above formula is applied to roof framing, we have,

A = the Rise

B = the Run

C= the Rafter length

Run = 1/2 building width - 1/2 ridge board width

Rise = H/12 x Run (H will be given in the plan; it is the amount of rise per foot of run)

<u>Rafter Cut Length = Rafter Length + Rafter Tail Length</u>

Finding Rafter Length:

First find the run and the rise using the above formula. Now, calculate the rafter length using this formula:

Rafter Length = $\sqrt{(\text{Rise} \times \text{Rise}) + (\text{Run} \times \text{Run})}$

Finding Rafter Tail Length:

The rafter tail length is calculated using the formula:

Rafter Tail Length = $\sqrt{(\text{TRise x TRise}) + (\text{TRun xTRun})}$

First calculate the TRun:

TRun = Overhang - Fascia

Now calculate TRise:

TRise = H/12 x TRun

Now, you can calculate the Rafter Tail Length and finally the Rafter Cut Length

Chalking Lines Duplication Method

If the Pythagoras theorem method is difficult, use Chalking Lines Duplication method, as it is the easiest method available to determine the rafter length and the rafter cut length.

In this method, you will draw the actual size of the rafter on your floor and measure its length. The details required for this method are all given in the plans. Following are the details that you need to keep ready:

A. Pitch

B Span

C Width of rafter

D Length of the Overhang

E Size of the exterior wall

Steps for Chalking Lines Duplication Method:

1. Draw a straight line using a chalk; the line must be longer than the rafter length

2. Draw two more lines perpendicular to the first line; these lines represent the exterior wall

3. Mark the point where the inside exterior wall line crosses the ceiling joist line. From this point, based on the pitch, measure out and measure up. If your pitch is 6 by 12, measure a horizontal distance of 12" and from the 12" point, measure a vertical distance of 6"

4. Chalk a line for the thickness of the rafter

Steps for Roof Framing

The steps mentioned in the above pages are for calculating the rafter cut length, which is the 1st step for framing. Following are the remaining steps to be followed to complete your roof framing (any type of framing):

Step 2: Cut Common Rafter

Step 3: Set Ridge board

Step 4: Set Common Rafters

Step 5: Calculate the length of hip and valley rafters

Step 6: Cut and set the hip and valley rafters

Step 7: Set jack rafters, block rafters, and lookouts

Step 8: Set Fascia

Framing a Gable Roof:

While the above are the general steps for any type of roof framing, the following steps are for specifically framing a Gable roof:

1. Before you start framing the roof, check if the ceiling joists are nailed in place
2. Lay a roof sheathing on top of the ceiling joists -- this will be your platform to frame the roof
3. Layout the ridge board; your ridge board must be one size bigger than the common rafters; if you are using 2x4 rafters, use 2x6 ridge board -- this ensures proper bearing
4. On the ridge board mark the gable end overhang
5. Lay the gable end rafter on the inside of the wall
6. Now start laying the common rafters
7. Secure the rafters using nails or screws
8. Finally install the fascia

Roof Sheathing

Once the roof framing is completed, the next step is to sheath the roof. The most important thing to remember when sheathing the roof is to provide space between the sheathing sheets; this is required to provide space for expansion, especially in mist climates.

- Provide 8d nails between to provide space for expansion.
- Before laying the first row of sheets, snap chalk lines on the rafters
- If you have a pitch greater than 8/12, place temporary stickers; it will help you in walking
- Start applying the sheets from the eaves and go up towards the ridge
- Use nails at 6" intervals at the edge and at 12" intervals at the other places
- Sheath one side of the roof at a time

Sheathing Mistakes :

- The seams of the sheathing are not staggered
- Not aligning the sheathing perpendicular to the rafters
- Using the sheathing wrong side up
- Not using enough nails/screws
- Panels do not meet at the center of the rafter

The next step in framing your roof is installing a roofing material of your choice. After this is done, install a siding of your choice on the roofs. Siding is offered in various materials. When you buy the siding material, consider the wind factor and the fact that you will travel on the highway -- make enough allowances for the same. Lap wood siding is also a good choice of siding; however, with this siding material, you must install lath. (Lath is a thin strip of wood used as a base for plaster).

Some people cover their tiny homes with Tyvek, as it makes the house watertight. You must remember that tiny houses have condensation problems and making it water tight will aggravate the condensation issues; instead, you can use felt/tar paper.

Chapter Nine: Doors and Windows

Door Installation

When installing doors and windows, read the instructions that came along with them very carefully. Double check the swing of the doors before installing. Most people use either a 32" or 36" door. The main concern with installing either doors or windows is that you must get them perfectly level.

Following are the steps for door installation:

1. Read the instructions carefully and check plans
2. Check the threshold for level; this establishes a level door leading to longevity
3. Nail hinge jamb (jamb is an upright member on the door frame); use a 16d galvanized casing nail at each hinge
4. Next, nail the latch jamb; before nailing, there must be continuous gap of 1/8" between the door and the jamb; the door must touch the latch jamb equally throughout
5. Check for the gap around the door - it must be even; when you close the door, it must not rattle
6. Using 16d galvanised casing nails, nail the door

Window Installation Steps:

The rough opening in your wall must be cut to dimensions mentioned in the instruction guide (it comes with the window). Place the window in the gap (you might require assistance for setting the window) and use shims if the window doesn't fit properly. You can check if the window is level using a spirit level.

Now take a piece of chalk and mark around the flanges. Cut into the siding to accommodate the flanges, you must remove the window before doing so.

Apply building paper to the window edges; measure the length of the edges, mark them on the paper and cut accordingly.

The next step is to add caulking -- it prevents mold, rot, and keeps away cold. Buy a caulk that is used for outside.

After applying the caulk, install the window into the opening and ensure that it is level.

Nail down the window -- start at the corner and nail at every 10".

Chapter Ten: A note on Wiring and Plumbing

Wiring

It is always a good idea to hire professional help for wiring and plumbing. A tiny house doesn't need elaborate electrical wiring or plumbing. So, if you are comfortable with doing your own wiring and plumbing, te following are a few pointers:

- Rough-in is the stage where all the wires, electrical boxes, circuit breakers, and other necessary electrical items are installed. Wires are pulled in through the joists, studs, and outlet boxes; small holes might be drilled to pull in the wires
- Mark all the points where electrical items will be located
- Walk through your house and check if the type and number of fixtures to be installed are correct
- Check for light switch placement
- Install circuit breakers
- Holes are drilled and wires are pulled in through the house frame
- Make sure that the wires are neat and organized; it will help you in troubleshooting problems that might arise in the future
- The final step is to cap all circuits and mark the wires

Plumbing

Similar to rough-in wiring, rough-in plumbing involves laying down the pipes for drains and water supply lines. If you are using a water heater, this is time to install it. Usually tanks for potable water and holding wastage are installed at the bottom of the trailer during floor framing. If you plan on using storage and holding tanks and have not yet done it, now is the time to install the tanks.

Most people use PVC pipes for plumbing work. Following are few other materials to consider for plumbing:

- Copper -- rigid and flexible (If you are using rigid Copper, solder tees and joints using a flux)
- PVC (rigid plastic)
- Galvanised steel
- Crosslink polyethylene

Recycling of Human waste

You can either use a holding tank for waste disposal or use the compost method. Following are some toilet options for THOW:

a. Composting toilets - these come in two types, low-tech and high-tech. A low tech composting toilet requires no plumbing and no sewage. It is a 5 gallon plastic bucket and has a toilet seat. You must use sawdust on the waste overtime and dump it outside on the compost pile. A high-tech compost toilets are self-contained toilets. They work with or without electricity; you can use the mechanical spurring action for stirring.
b. Low-flush toilets - these toilets come with a holding tank; you must empty the holding tank into a sewage system or some other place.
c. Incinerating toilets - these use 20amps of power to burn waste; you can empty the ash into a trash

Chapter Eleven: Tips

Helpful Tips

Transitioning to a tiny house from a big house is not easy and can be challenging. Before you decide to move into a tiny house, the following are a few pointers:

a) How much space you need - this decision must be based on what you **need** and not on how much stuff you have.

b) How many people will live in your tiny house - if you are single, you have the freedom to choose a place as tiny as you want; however, if there are others living with you, you must take their space needs into consideration too.

c) If you are a couple, take into consideration that you might need an area to retreat.

d) If you have kids, building a separate room for them is always not necessary; consider using bunk beds and other creative space saving solutions.

e) If you entertain guests and/or work from home, make leeway for the same.

f) You will definitely have to downsize your wardrobe, which is generally a major task.

g) It will take time for you to adjust with the tiny space for the countertops in the kitchen; prepping food will also consume more time (until you get adjusted to the new space).

h) Initially, if there are people living with you, you might bump into them; once you settle down, you will thank the tiny space as it fosters more proximity among the members.

i) You will also cultivate the habit of putting back everything into its place after using them.

Pointers for decluttering your wardrobe:

i. Give away clothes that you haven't worn in the past year

ii. Separate out clothes that don't fit properly

iii. You might have clothes that you want to wear but are unsuitable for the climate; use vacuum bags to store such clothing

iv. You will also have clothes that you plan to get fixed (you have been planning for a long time now) -- toss those clothes

v. If you are unsure about what to do with certain items, pack them and check them again in a couple of months; chances are that you will decide to get rid of them

Plans

House #1: 325 ft^2

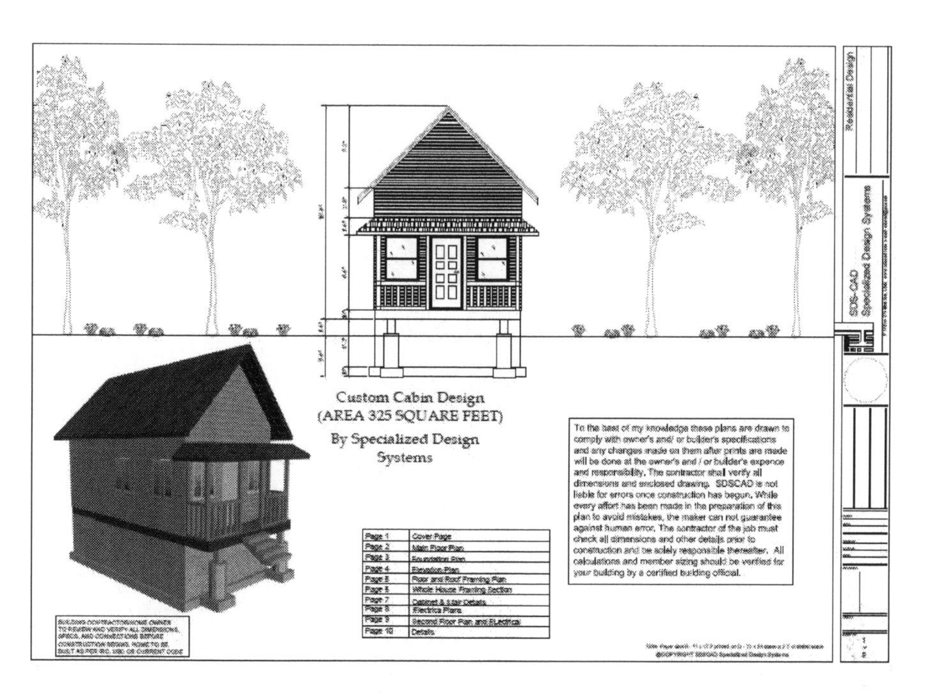
Custom Cabin Design
(AREA 325 SQUARE FEET)
By Specialized Design Systems
To the best of my knowledge these plans are drawn to comply with owner's and/ or builder's specifications and any changes made on them after prints are made will be done at the owner's and / or builder's expence and responsibility. The contractor shal verify all dimensions and enclosed drawing. SDSCAD is not liable for errors once construction has begun. While every effort has been made in the preparation of this plan to avoid mistakes, the maker can not guarantee against human error. The contractor of the job must check all dimensions and other details prior to construction and be solely responsible thereafter. All calculations and member sizing should be verified for your building by a certified building official.
Page 1 Cover Page
Page 2 Main Floor Plan
Page 3 Foundation Plan
Page 4 Elevation Plan
Page 5 Floor and Roof Framing Plan
Page 6 Whole House Framing Section
Page 7 Cabinet & Stair Details
Page 8 Electrical Plans
Page 9 Second Floor Plan and Electrical
Page 10 Details
Residential Design
SDS-CAD
Specialized Design Systems

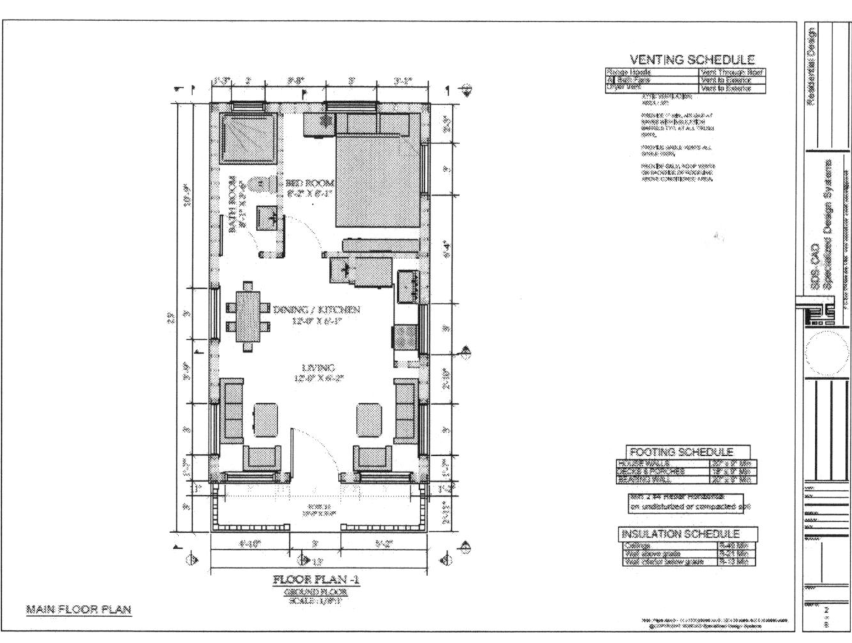
VENTING SCHEDULE
FOOTING SCHEDULE
INSULATION SCHEDULE
BATH ROOM
BED ROOM
DINING / KITCHEN
LIVING
FLOOR PLAN -1
GROUND FLOOR
MAIN FLOOR PLAN
Residential Design
SDS-CAD
Specialized Design Systems

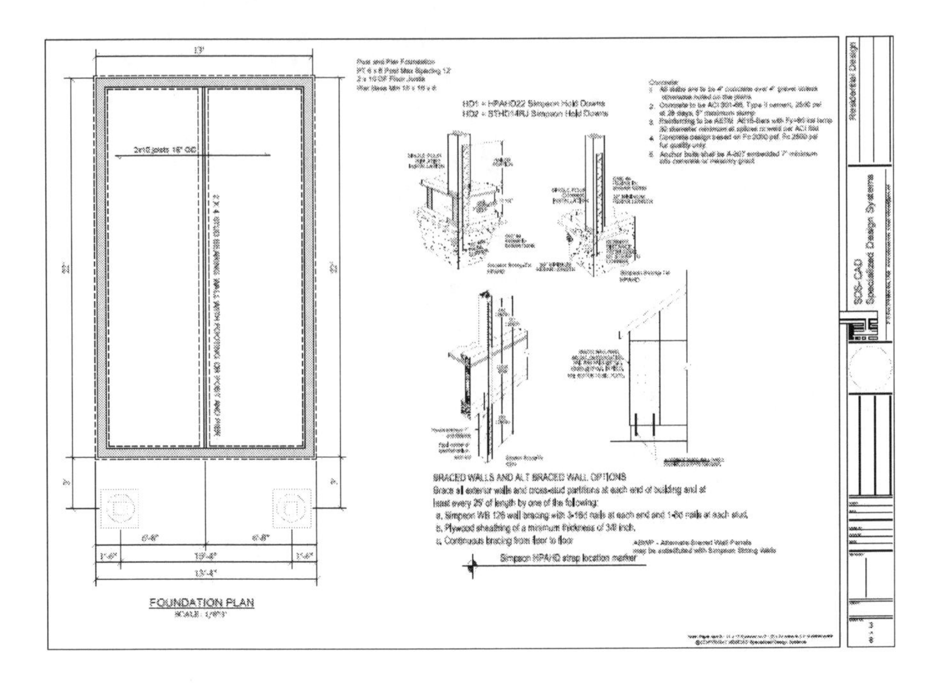
13'
2x10 joists 16" OC
HD1 = HPAHD22 Simpson Hold Downs
HD2 = STHD14RJ Simpson Hold Downs
BRACED WALLS AND ALT BRACED WALL OPTIONS
Brace all exterior walls and cross-stud partitions at each end of building and at
least every 25' of length by one of the following:
a, Simpson WB 126 wall bracing with 3-16d nails at each end and 1-8d nails at each stud,
b, Plywood sheathing of a minimum thickness of 3/8 inch,
c, Continuous bracing from floor to floor
Simpson HPAHD strap location marker
6'-8"
6'-8"
1'-6"
10'-4"
1'-6"
13'-4"
FOUNDATION PLAN
Residential Design
SDS-CAD
Specialized Design Systems

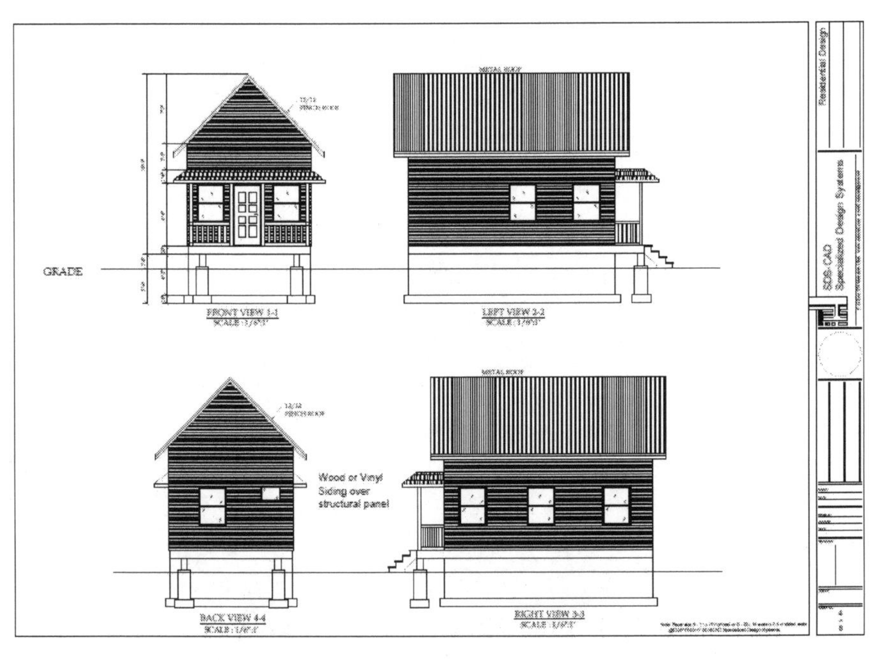
METAL ROOF
GRADE
FRONT VIEW 1-1
LEFT VIEW 2-2
METAL ROOF
Wood or Vinyl
Siding over
structural panel
BACK VIEW 4-4
RIGHT VIEW 3-3
Residential Design
SDS-CAD
Specialized Design Systems

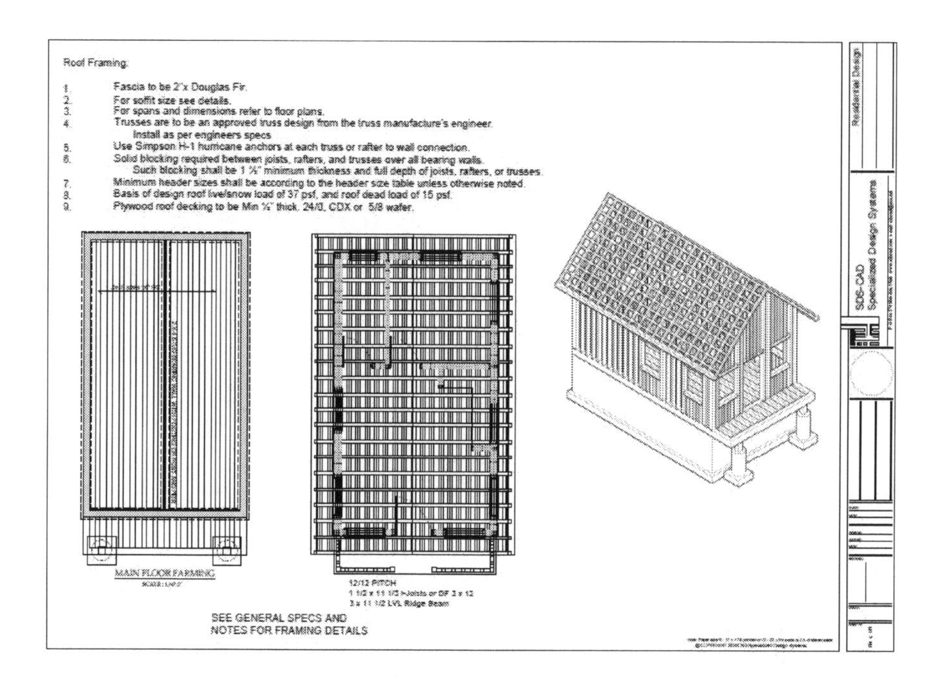
Roof Framing:
1. Fascia to be 2"x Douglas Fir.
2. For soffit size see details.
3. For spans and dimensions refer to floor plans.
4. Trusses are to be an approved truss design from the truss manufacture's engineer. Install as per engineers specs
5. Use Simpson H-1 hurricane anchors at each truss or rafter to wall connection.
6. Solid blocking required between joists, rafters, and trusses over all bearing walls. Such blocking shall be 1 ½" minimum thickness and full depth of joists, rafters, or trusses.
7. Minimum header sizes shall be according to the header size table unless otherwise noted.
8. Basis of design roof live/snow load of 37 psf, and roof dead load of 15 psf.
9. Plywood roof decking to be Min ½" thick. 24/0, CDX or 5/8 wafer.
MAIN FLOOR FARMING
12/12 PITCH
SEE GENERAL SPECS AND
NOTES FOR FRAMING DETAILS
Residential Design
SDS-CAD
Specialized Design Systems

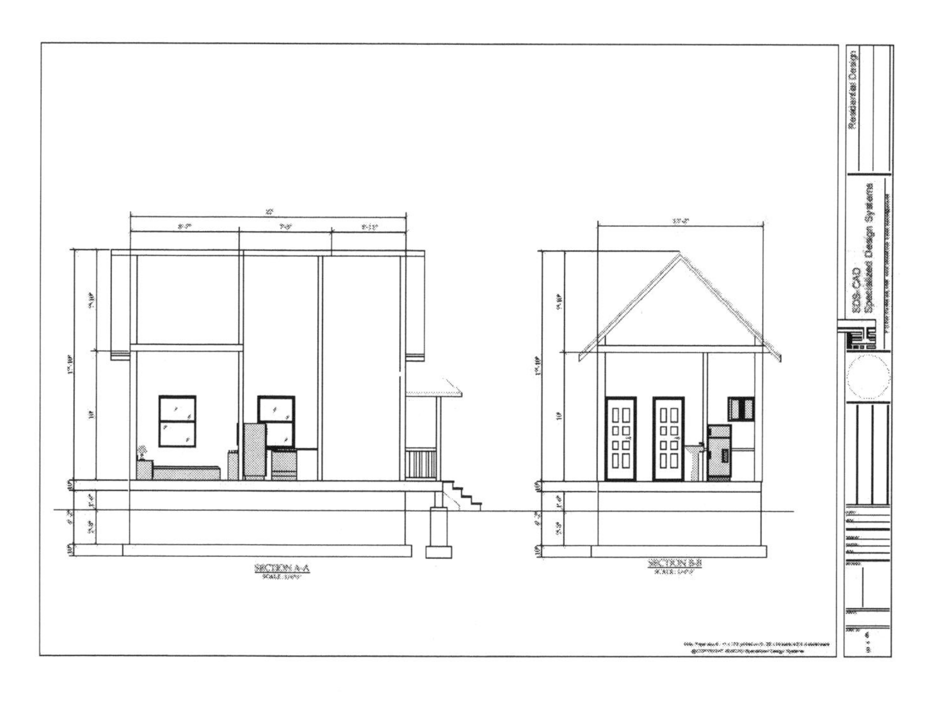
SECTION A-A
SECTION B-B
Residential Design
SDS-CAD
Specialized Design Systems

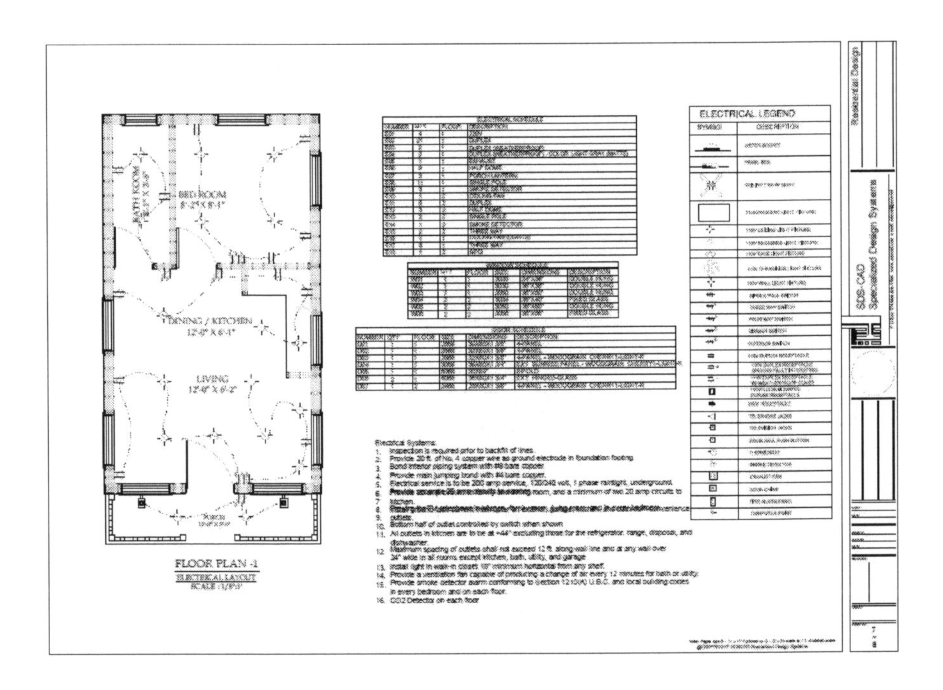

FLOOR PLAN -1
ELECTRICAL LAYOUT
ELECTRICAL LEGEND
SYMBOL
DESCRIPTION
Residential Design
SDS-CAD
Specialized Design Systems

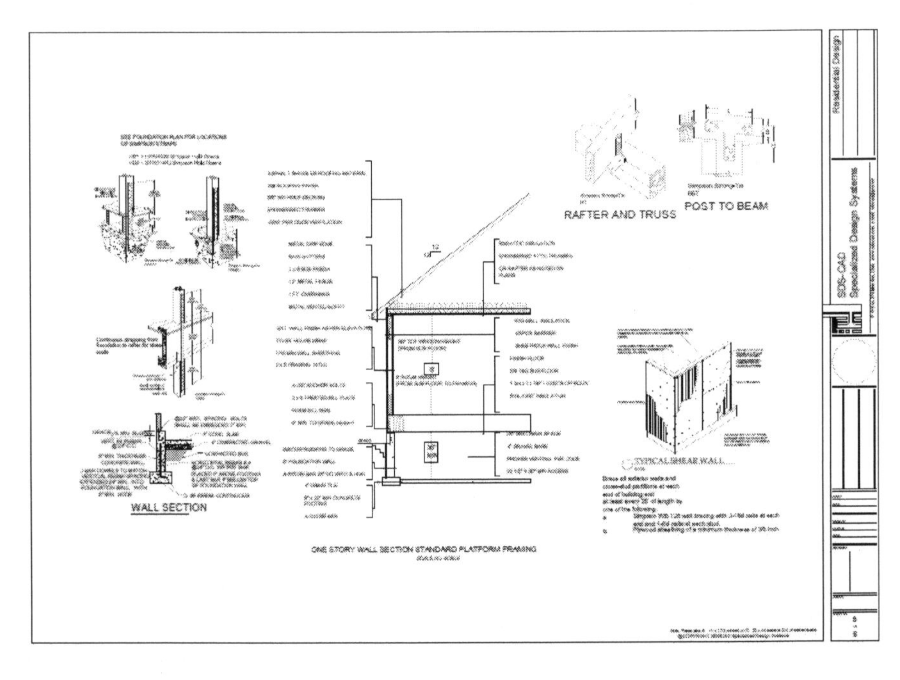

RAFTER AND TRUSS
POST TO BEAM
WALL SECTION
TYPICAL SHEAR WALL
ONE STORY WALL SECTION STANDARD PLATFORM FRAMING
Residential Design
SDS-CAD
Specialized Design Systems

House #2: 160 ft^2

Custom Cabin Design
(AREA 160 SQUARE FEET)
By Specialized Design
Systems
To the best of my knowledge these plans are drawn to comply with owner's and/ or builder's specifications and any changes made on them after prints are made will be done at the owner's and / or builder's expence and responsibility. The contractor shall verify all dimensions and enclosed drawing. SDSCAD is not liable for errors once construction has begun. While every effort has been made in the preparation of this plan to avoid mistakes, the maker can not guarantee against human error. The contractor of the job must check all dimensions and other details prior to construction and be solely responsible thereafter. All calculations and member sizing should be verified for your building by a certified building official.
Page 1 Cover Page
Page 2 Main Floor Plan
Page 3 Foundation Plan
Page 4 Elevation Plan
Page 5 Floor and Roof Framing Plan
Page 6 Whole House Framing Section
Page 7 Cabinet & Stair Details
Page 8 Electrical Plans
Page 9 Second Floor Plan and Electrical
Page 10 Details
Residential Design
SDS-CAD
Specialized Design Systems

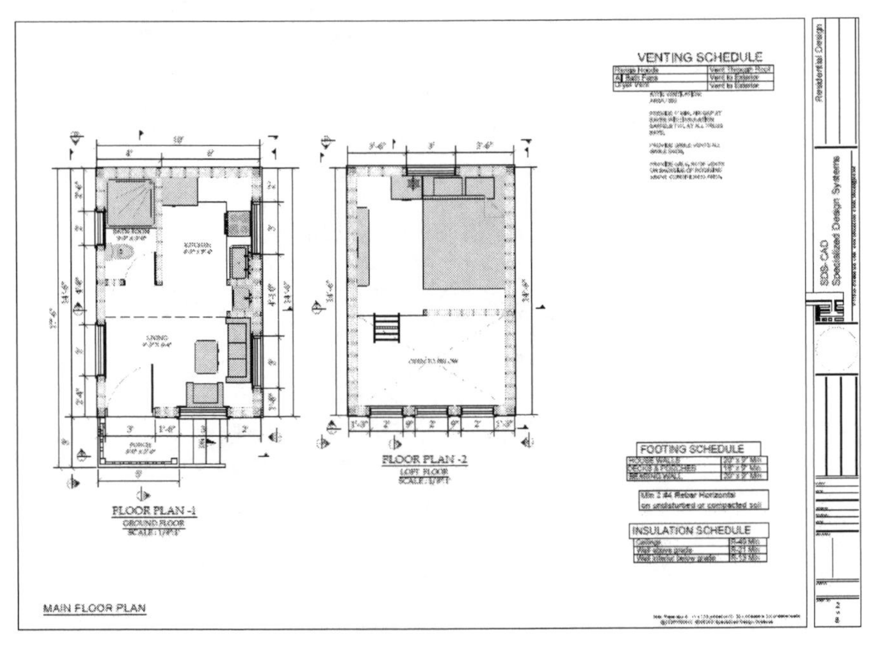
VENTING SCHEDULE
Range Hoods Vent Through Roof
All Bath Fans Vent to Exterior
Dryer Vent Vent to Exterior
FLOOR PLAN -1
GROUND FLOOR
FLOOR PLAN -2
LOFT FLOOR
FOOTING SCHEDULE
INSULATION SCHEDULE
MAIN FLOOR PLAN
Residential Design
SDS-CAD
Specialized Design Systems

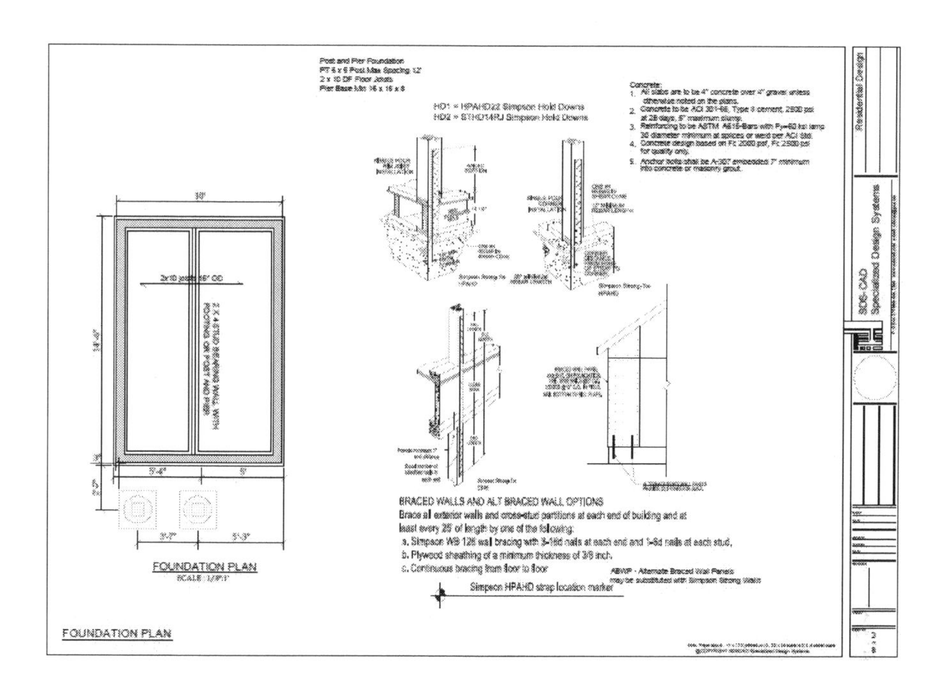
Post and Pier Foundation
PT 6 x 6 Post Max Spacing 12'
2 x 10 DF Floor Joists
HD1 = HPAHD22 Simpson Hold Downs
HD2 = STHD14RJ Simpson Hold Downs
BRACED WALLS AND ALT BRACED WALL OPTIONS
Brace all exterior walls and cross-stud partitions at each end of building and at least every 25' of length by one of the following:
b. Plywood sheathing of a minimum thickness of 3/8 inch.
c. Continuous bracing from floor to floor
Simpson HPAHD strap location marker
FOUNDATION PLAN
FOUNDATION PLAN
Residential Design
SDS-CAD
Specialized Design Systems

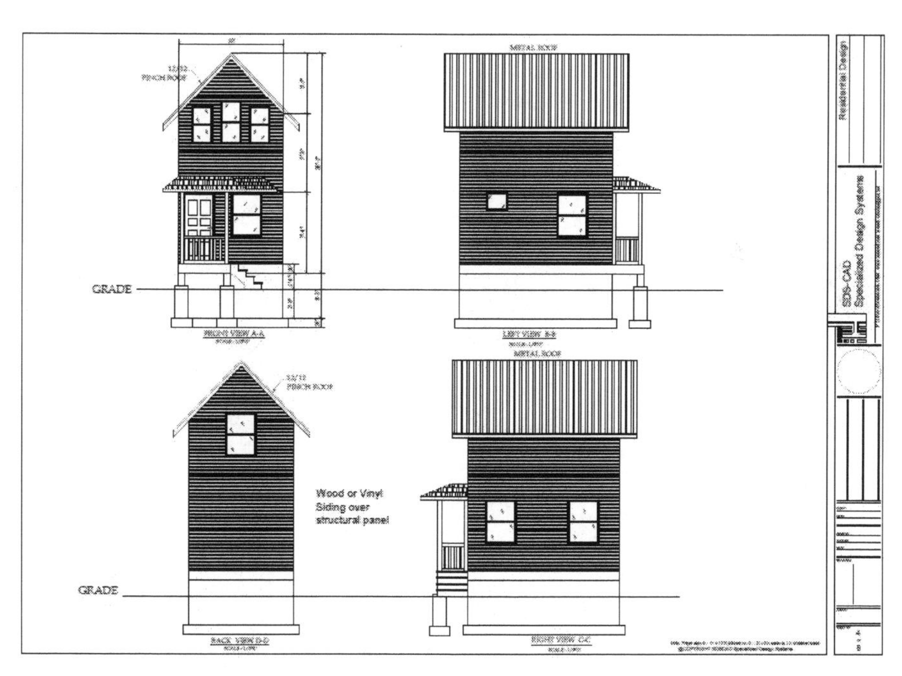
METAL ROOF
GRADE
METAL ROOF
Wood or Vinyl
Siding over
structural panel
GRADE
Residential Design
SDS-CAD
Specialized Design Systems

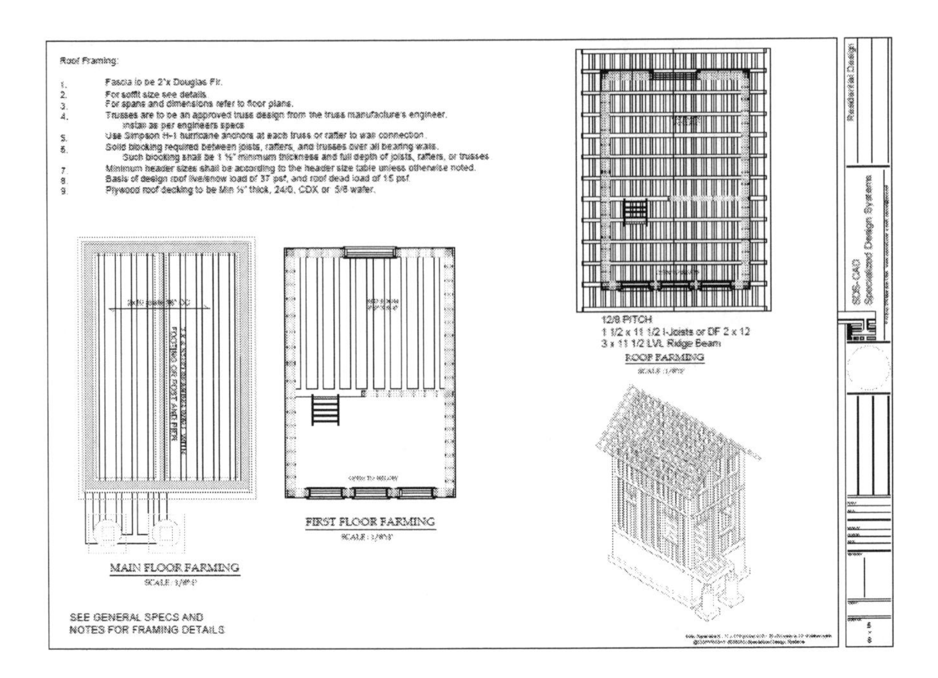

Roof Framing:
1. Fascia to be 2"x Douglas Fir.
2. For soffit size see details.
3. For spans and dimensions refer to floor plans.
4. Trusses are to be an approved truss design from the truss manufacture's engineer. install as per engineers specs
5. Use Simpson H-1 hurricane anchors at each truss or rafter to wall connection.
6. Solid blocking required between joists, rafters, and trusses over all bearing walls. Such blocking shall be 1 ½" minimum thickness and full depth of joists, rafters, or trusses
7. Minimum header sizes shall be according to the header size table unless otherwise noted.
8. Basis of design roof live/snow load of 37 psf, and roof dead load of 15 psf.
9. Plywood roof decking to be Min ½" thick, 24/0, CDX or 5/8 wafer.
12/8 PITCH
1 1/2 x 11 1/2 I-Joists or DF 2 x 12
3 x 11 1/2 LVL Ridge Beam
ROOF FARMING
FIRST FLOOR FARMING
MAIN FLOOR FARMING
SEE GENERAL SPECS AND
NOTES FOR FRAMING DETAILS
Residential Design
SDS-CAD
Specialized Design Systems

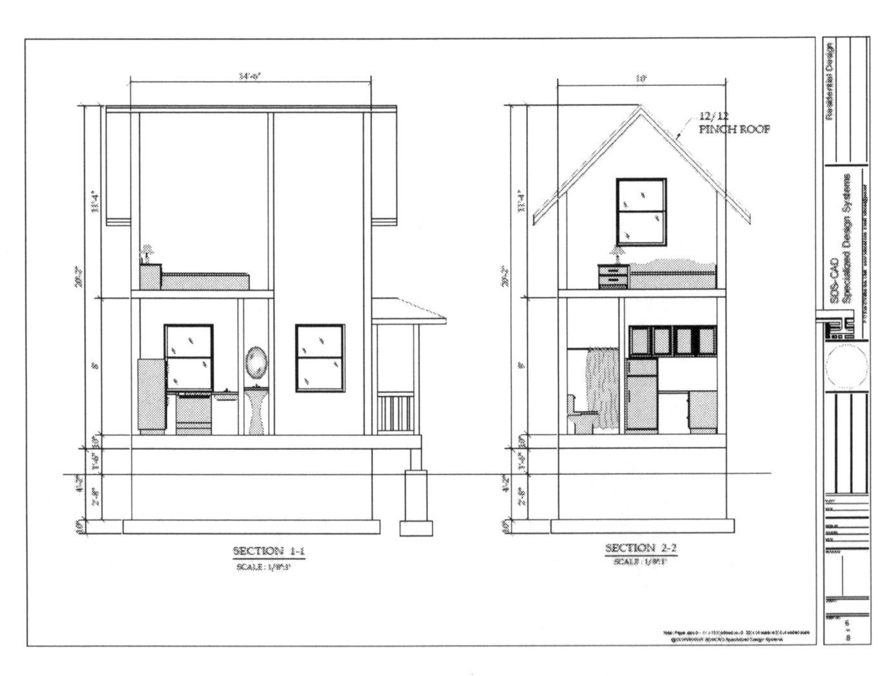

12/12
PINCH ROOF
SECTION 1-1
SECTION 2-2
Residential Design
SDS-CAD
Specialized Design Systems

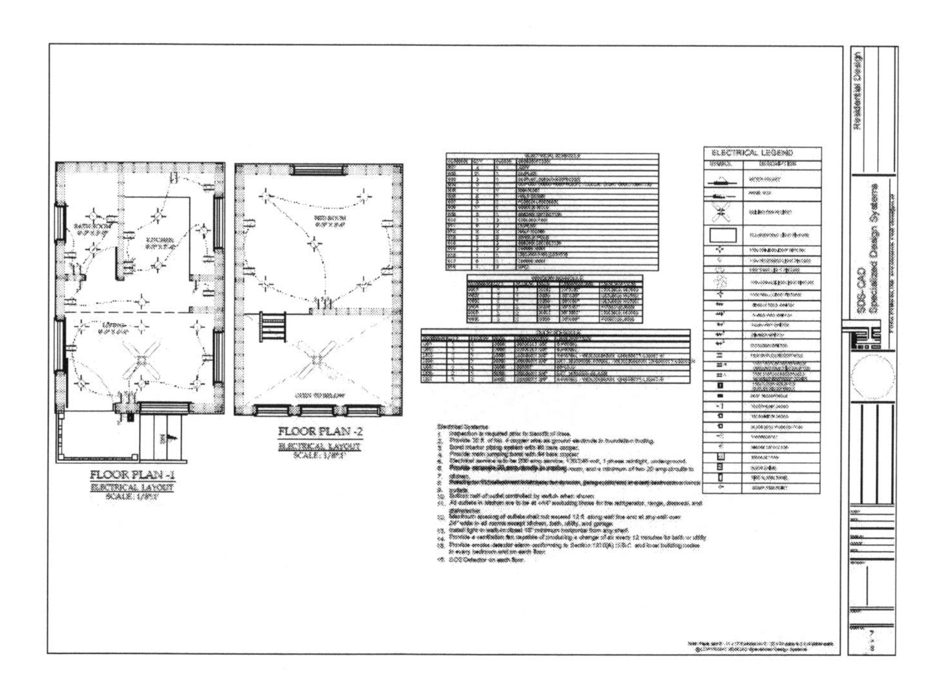
FLOOR PLAN -1
ELECTRICAL LAYOUT
FLOOR PLAN -2
ELECTRICAL LAYOUT
ELECTRICAL LEGEND
Residential Design
SDS-CAD
Specialized Design Systems

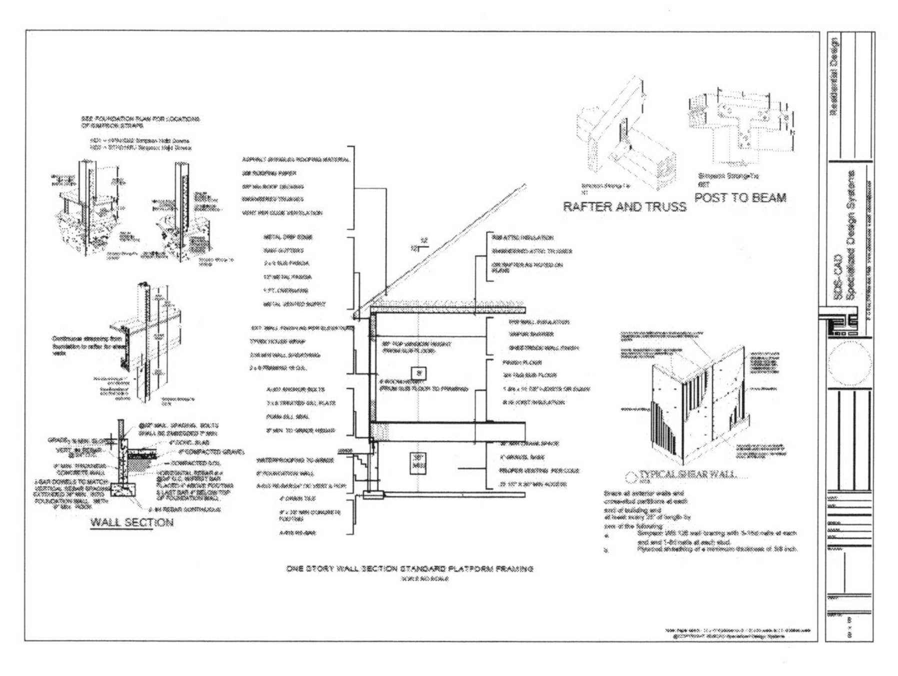
RAFTER AND TRUSS
POST TO BEAM
WALL SECTION
TYPICAL SHEAR WALL
ONE STORY WALL SECTION STANDARD PLATFORM FRAMING
Residential Design
SDS-CAD
Specialized Design Systems

House #3: 208 ft^2

Custom Cabin Design
(AREA 208 SQUIRE FEET)
By Specialized Design Systems
BUILDING CONTRACTOR/HOME OWNER TO REVIEW AND VERIFY ALL DIMENSIONS, SPECS, AND CONNECTIONS BEFORE CONSTRUCTION BEGINS, HOME TO BE BUILT AS PER IRC, UBC OR CURRENT CODE
Residential Design
SDS-CAD
Specialized Design Systems

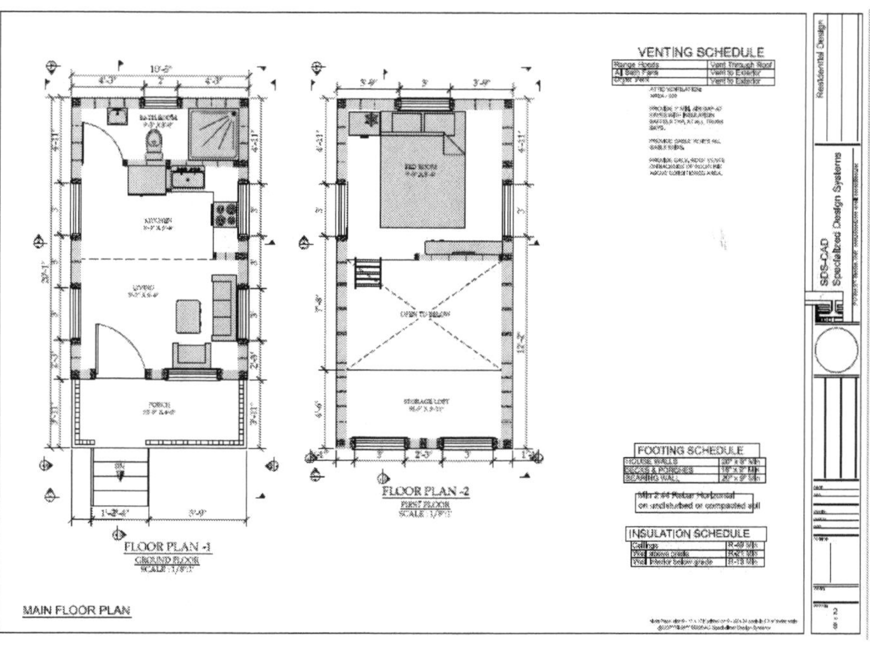
VENTING SCHEDULE
FLOOR PLAN -1
GROUND FLOOR
FLOOR PLAN -2
FIRST FLOOR
FOOTING SCHEDULE
INSULATION SCHEDULE
MAIN FLOOR PLAN
Residential Design
SDS-CAD
Specialized Design Systems

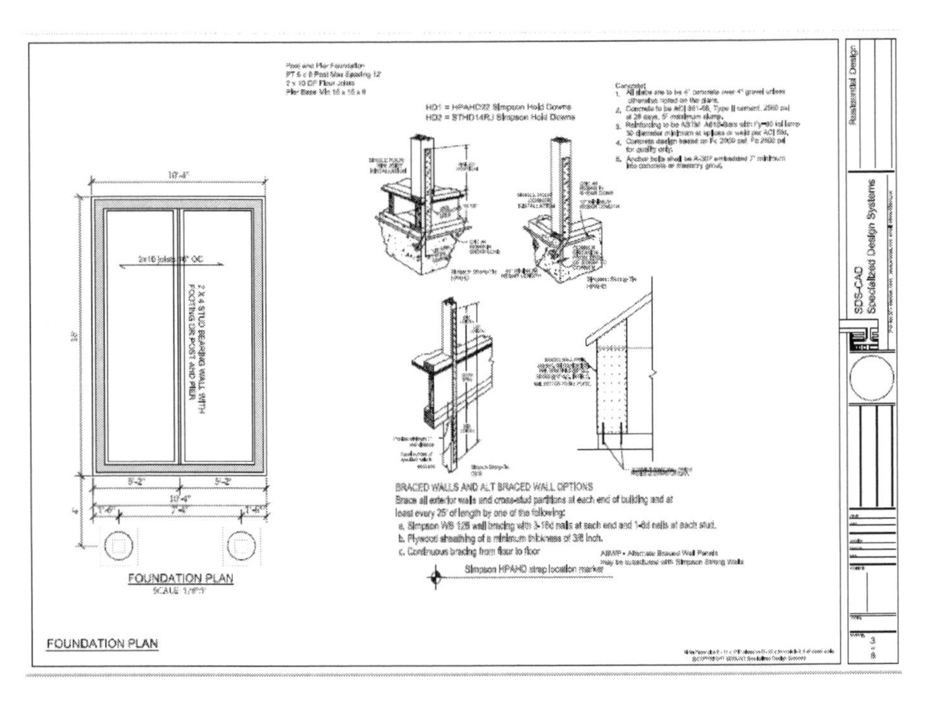
FOUNDATION PLAN
SCALE 1/4"=1'
BRACED WALLS AND ALT BRACED WALL OPTIONS
Brace all exterior walls and cross-stud partitions at each end of building and at least every 25' of length by one of the following:
a. Simpson WB 126 wall bracing with 3-16d nails at each end and 1-8d nails at each stud.
b. Plywood sheathing of a minimum thickness of 3/8 inch.
c. Continuous bracing from floor to floor
Simpson HPAHD strap location marker
HD1 = HPAHD22 Simpson Hold Downs
HD2 = STHD14RJ Simpson Hold Downs
2 X 4 STUD BEARING WALL WITH FOOTING OR POST AND PIER
Residential Design
SDS-CAD
Specialized Design Systems

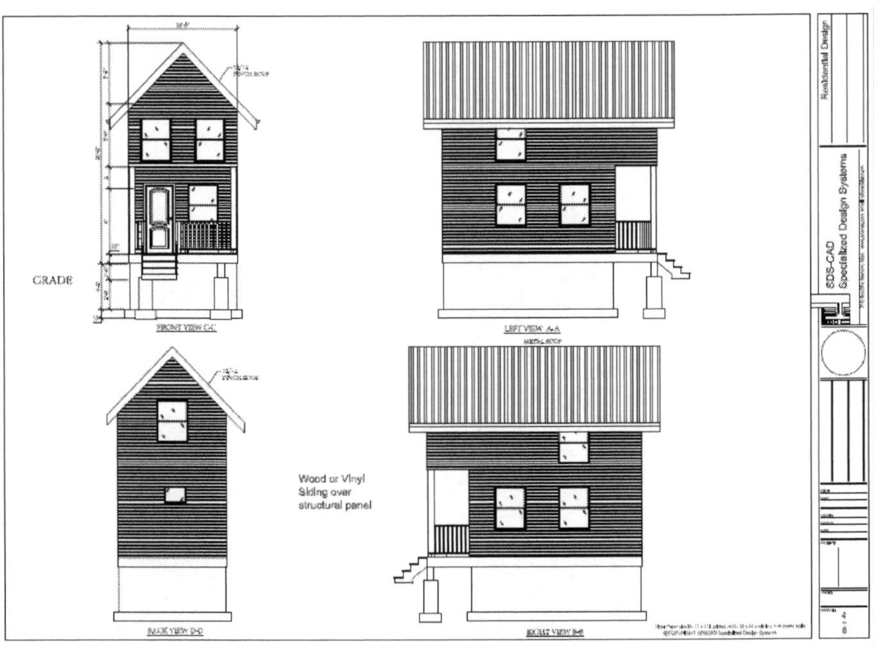
GRADE
Wood or Vinyl Siding over structural panel
Residential Design
SDS-CAD
Specialized Design Systems

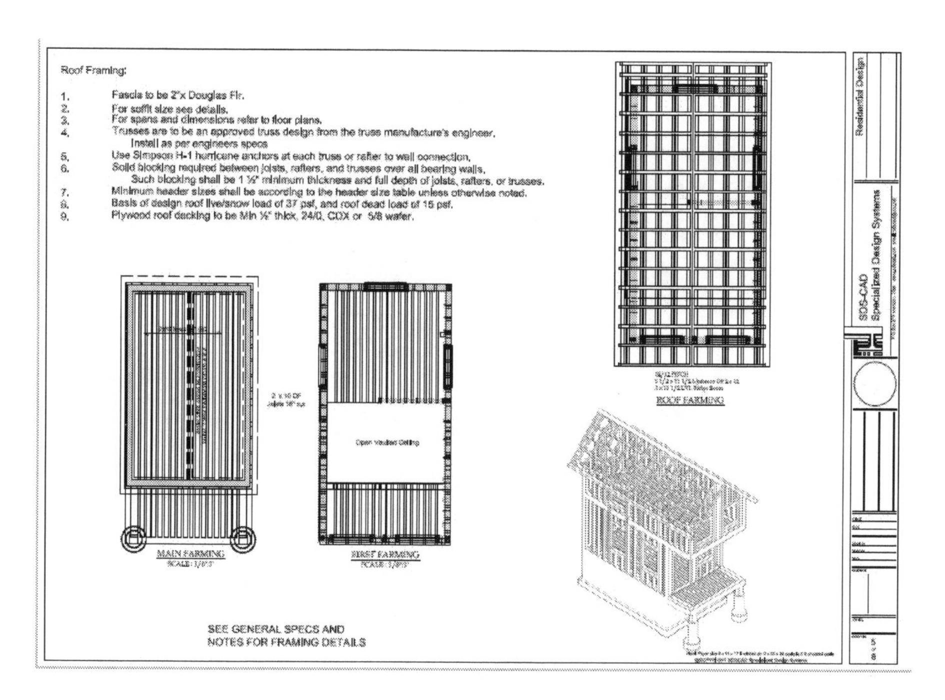

Roof Framing:
1. Fascia to be 2"x Douglas Fir.
2. For soffit size see details.
3. For spans and dimensions refer to floor plans.
4. Trusses are to be an approved truss design from the truss manufacture's engineer. Install as per engineers specs
5. Use Simpson H-1 hurricane anchors at each truss or rafter to wall connection.
6. Solid blocking required between joists, rafters, and trusses over all bearing walls. Such blocking shall be 1 ½" minimum thickness and full depth of joists, rafters, or trusses.
7. Minimum header sizes shall be according to the header size table unless otherwise noted.
8. Basis of design roof live/snow load of 37 psf, and roof dead load of 15 psf.
9. Plywood roof decking to be Min ½" thick, 24/0, CDX or 5/8 wafer.
Open Vaulted Ceiling
MAIN FARMING
FIRST FARMING
ROOF FARMING
SEE GENERAL SPECS AND
NOTES FOR FRAMING DETAILS
Residential Design
SDS-CAD
Specialized Design Systems

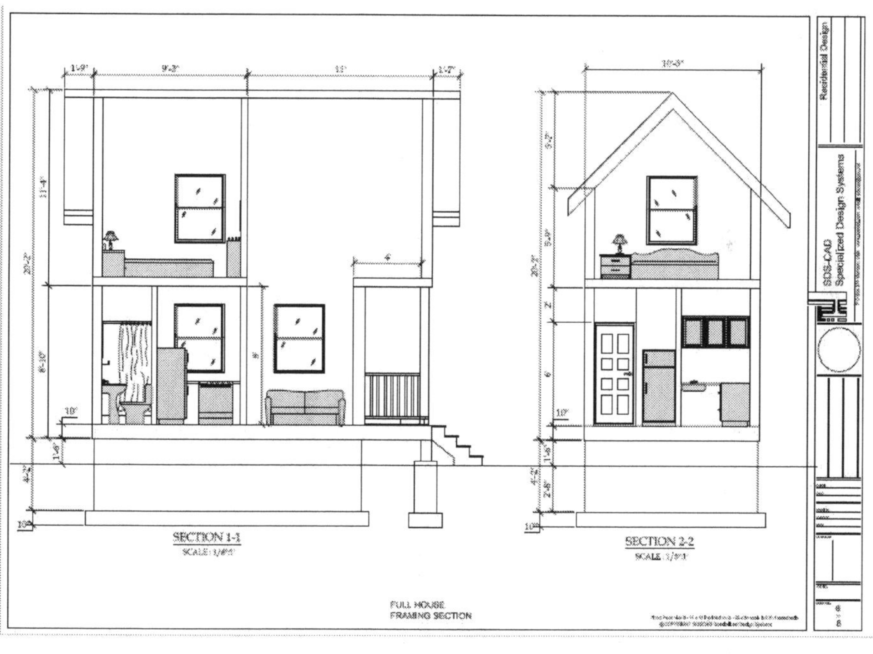

SECTION 1-1
SECTION 2-2
FULL HOUSE
FRAMING SECTION
Residential Design
SDS-CAD
Specialized Design Systems

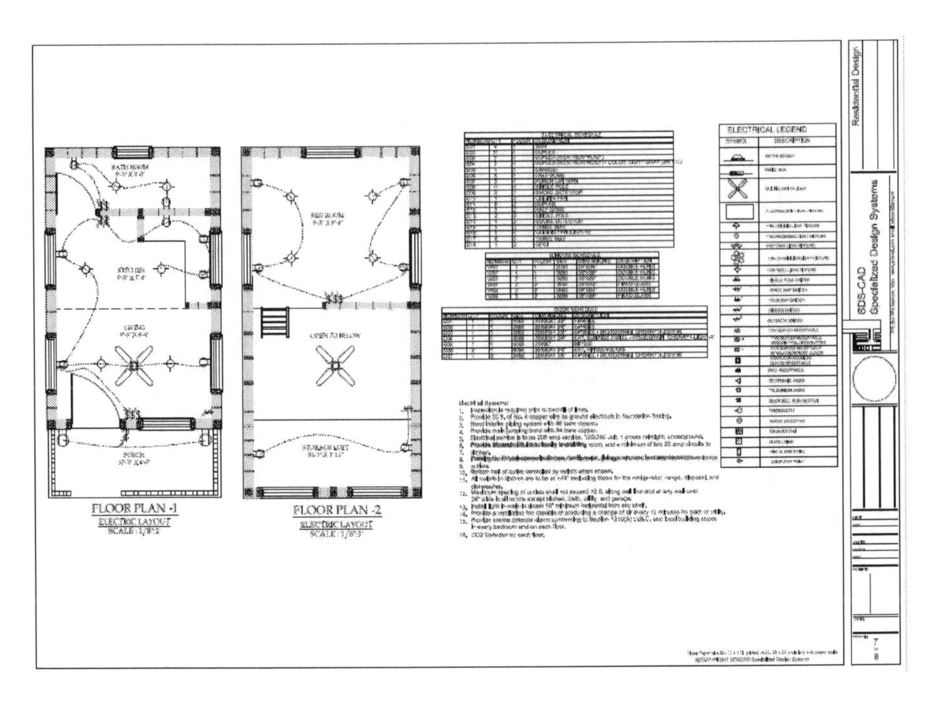

FLOOR PLAN -1
ELECTRIC LAYOUT
FLOOR PLAN -2
ELECTRIC LAYOUT
ELECTRICAL LEGEND
Residential Design
SDS-CAD
Specialized Design Systems

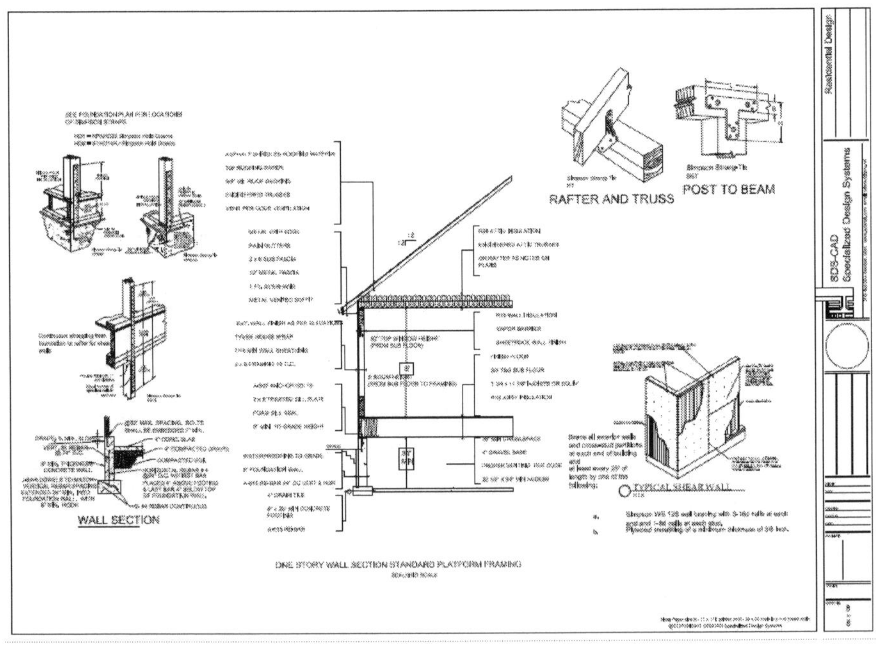

RAFTER AND TRUSS
POST TO BEAM
WALL SECTION
TYPICAL SHEAR WALL
ONE STORY WALL SECTION STANDARD PLATFORM FRAMING
Residential Design
SDS-CAD
Specialized Design Systems

House #4: 240 ft^2

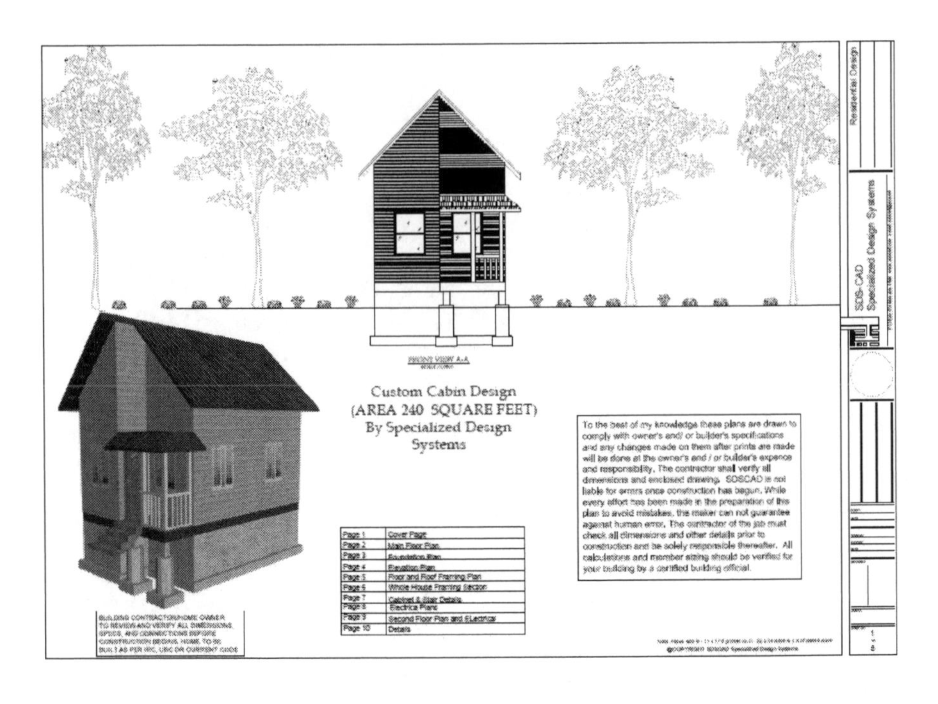
Custom Cabin Design
(AREA 240 SQUARE FEET)
By Specialized Design
Systems
To the best of my knowledge these plans are drawn to comply with owner's and/ or builder's specifications and any changes made on them after prints are made will be done at the owner's and / or builder's expence and responsibility. The contractor shall verify all dimensions and enclosed drawing. SDSCAD is not liable for errors once construction has begun. While every effort has been made in the preparation of this plan to avoid mistakes, the maker can not guarantee against human error. The contractor of the job must check all dimensions and other details prior to construction and be solely responsible thereafter. All calculations and member sizing should be verified for your building by a certified building official.
Page 1 Cover Page
Page 2 Main Floor Plan
Page 3 Foundation Plan
Page 4 Elevation Plan
Page 5 Floor and Roof Framing Plan
Page 6 Whole House Framing Section
Page 7 Cabinet & Stair Details
Page 8 Electrical Plans
Page 9 Second Floor Plan and Electrical
Page 10 Details
Residential Design
SDS-CAD
Specialized Design Systems

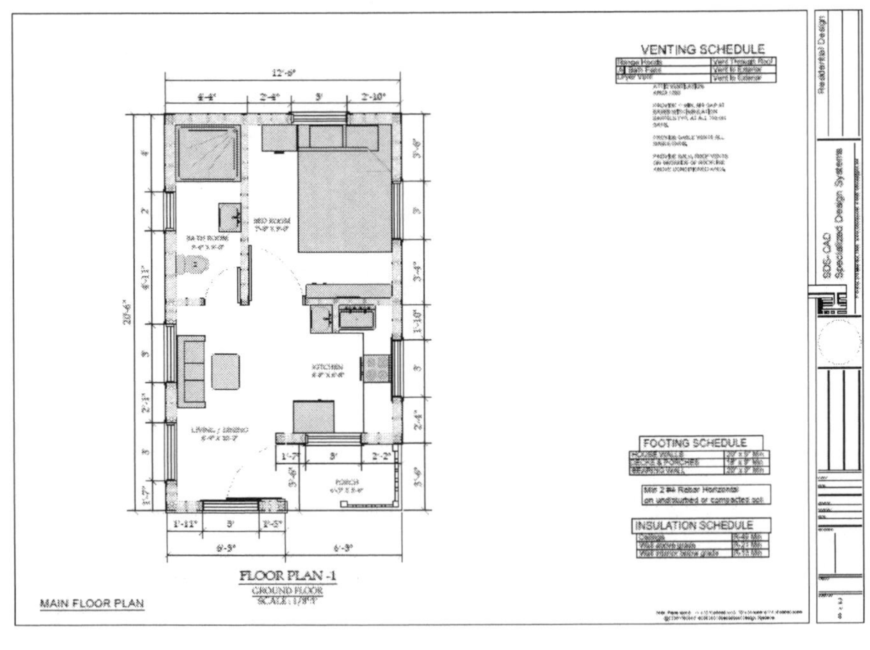
VENTING SCHEDULE
Range Hoods | Vent Through Roof
Vent to Exterior
Dryer Vent | Vent to Exterior
12'-6"
4'-4"
2'-4"
3'
2'-10"
20'-6"
BED ROOM
BATH ROOM
KITCHEN
LIVING / DINING
PORCH
1'-7"
2'-2"
1'-11"
1'-5"
6'-3"
6'-3"
FOOTING SCHEDULE
Min 2 #4 Rebar Horizontal
on undisturbed or compacted soil
INSULATION SCHEDULE
MAIN FLOOR PLAN
FLOOR PLAN -1
GROUND FLOOR
Residential Design
SDS-CAD
Specialized Design Systems

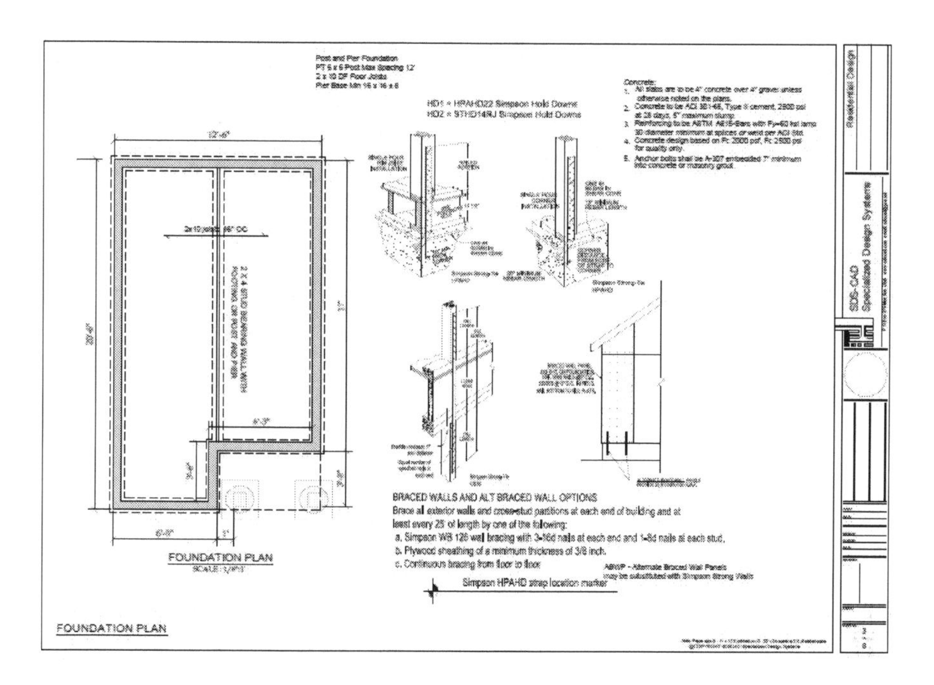
Post and Pier Foundation
PT 6 x 6 Post Max Spacing 12'
2 x 10 DF Floor Joists
Pier Base Min 16 x 16 x 8
HD1 = HPAHD22 Simpson Hold Downs
HD2 = STHD14RJ Simpson Hold Downs
Concrete:
1. All slabs are to be 4" concrete over 4" gravel unless otherwise noted on the plans.
2. Concrete to be ACI 301-65, Type II cement, 2500 psi at 28 days, 5" maximum slump.
3. Reinforcing to be ASTM A615-Bars with Fy=60 ksi lamp 30 diameter minimum at splices or weld per ACI Std.
4. Concrete design based on Fc 2500 psf, Fc 2500 psi for quality only.
5. Anchor bolts shall be A-307 embedded 7" minimum into concrete or masonry grout.
12'-6"
20'-6"
2x10 joists 16" OC
2 X 4 STUD BEARING WALL WITH FOOTING OR POST AND PIER
6'-3"
FOUNDATION PLAN
BRACED WALLS AND ALT BRACED WALL OPTIONS
Brace all exterior walls and cross-stud partitions at each end of building and at least every 25' of length by one of the following:
a. Simpson WB 126 wall bracing with 3-16d nails at each end and 1-8d nails at each stud.
b. Plywood sheathing of a minimum thickness of 3/8 inch.
c. Continuous bracing from floor to floor
ABWP - Alternate Braced Wall Panels may be substituted with Simpson Strong Walls
Simpson HPAHD strap location marker
FOUNDATION PLAN
Residential Design
SDS-CAD
Specialized Design Systems

GRADE
Wood or Vinyl
Siding over
structural panel
GRADE
Residential Design
SDS-CAD
Specialized Design Systems

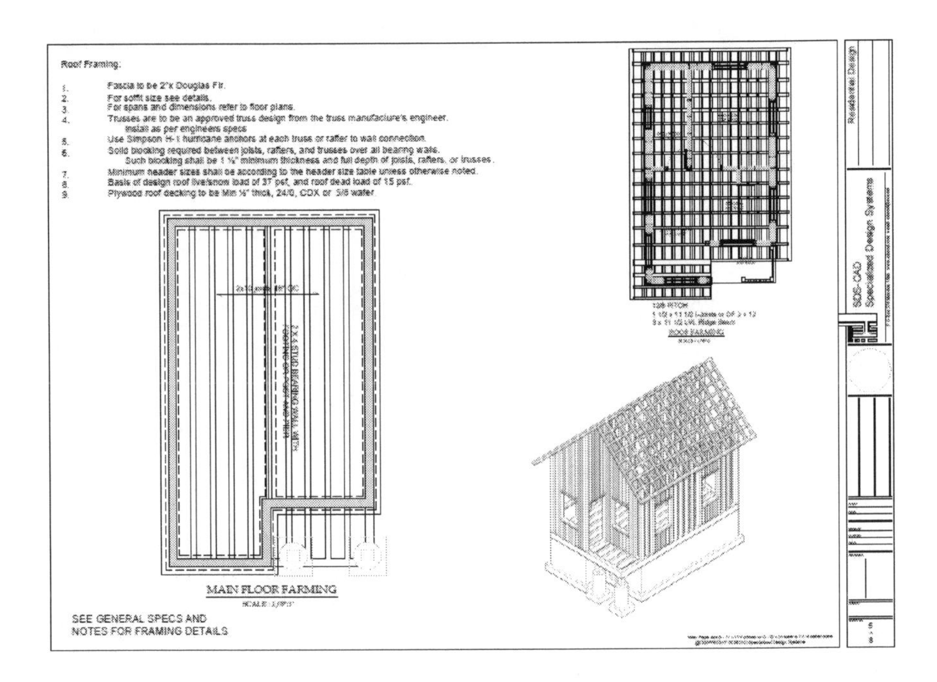
Roof Framing:
1. Fascia to be 2"x Douglas Fir.
2. For soffit size see details.
3. For spans and dimensions refer to floor plans.
4. Trusses are to be an approved truss design from the truss manufacture's engineer. Install as per engineers specs
5. Use Simpson H-1 hurricane anchors at each truss or rafter to wall connection.
6. Solid blocking required between joists, rafters, and trusses over all bearing walls. Such blocking shall be 1 ½" minimum thickness and full depth of joists, rafters, or trusses.
7. Minimum header sizes shall be according to the header size table unless otherwise noted.
8. Basis of design roof live/snow load of 37 psf, and roof dead load of 15 psf.
9. Plywood roof decking to be Min ½" thick, 24/0, CDX or 5/8 wafer.
MAIN FLOOR FARMING
SEE GENERAL SPECS AND
NOTES FOR FRAMING DETAILS
ROOF FARMING
Residential Design
SDS-CAD
Specialized Design Systems

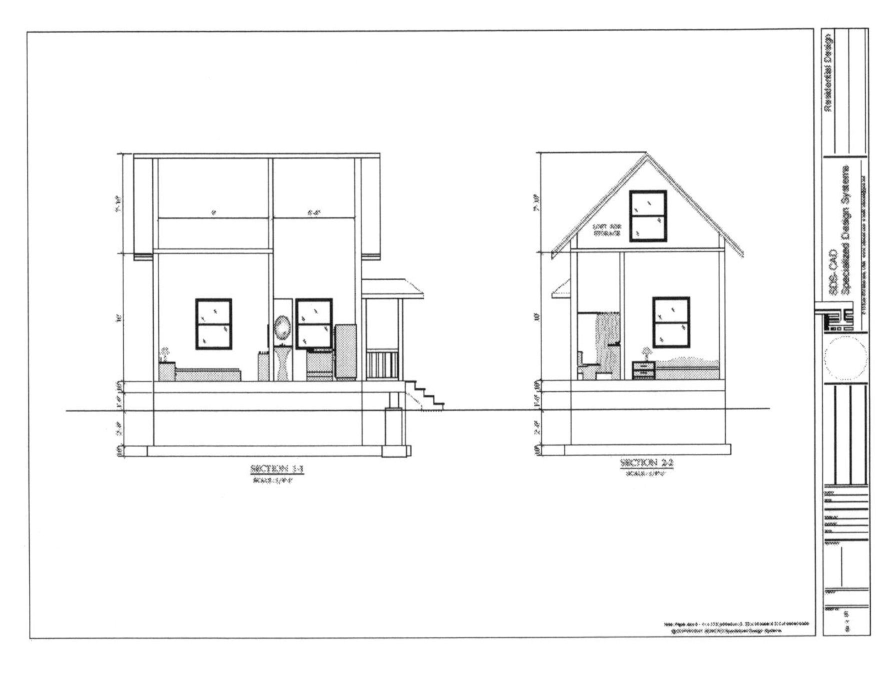
SECTION 1-1
SECTION 2-2
Residential Design
SDS-CAD
Specialized Design Systems

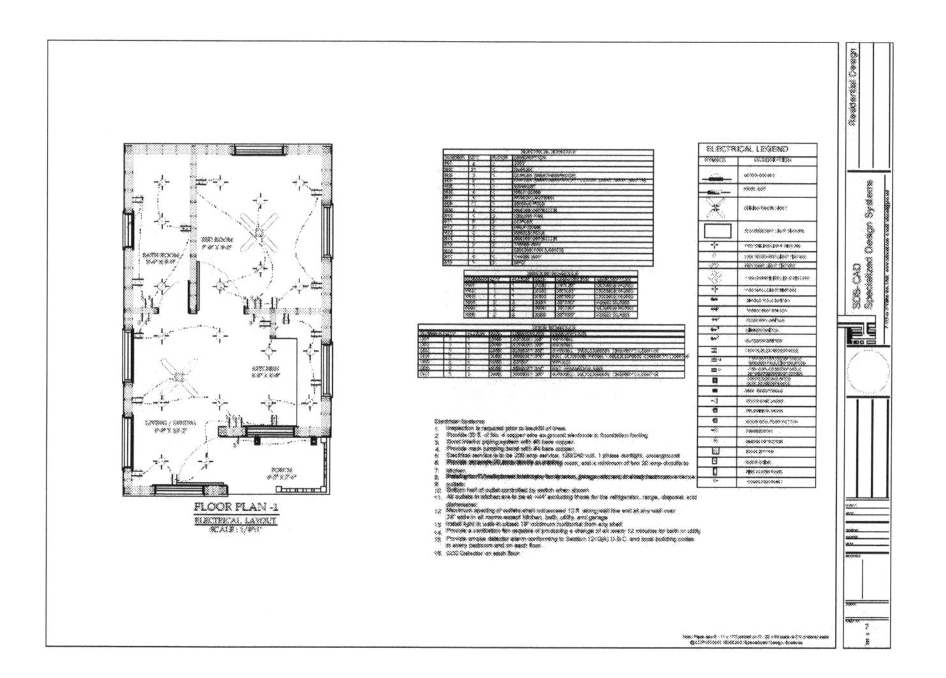
FLOOR PLAN -1
ELECTRICAL LAYOUT
ELECTRICAL LEGEND
Residential Design
SDS-CAD
Specialized Design Systems

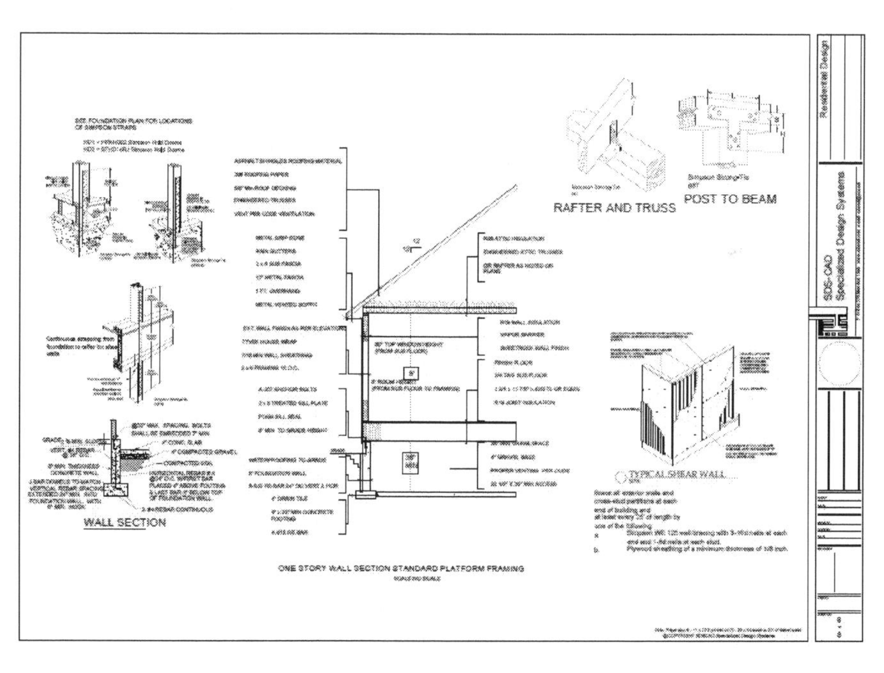
RAFTER AND TRUSS
POST TO BEAM
WALL SECTION
TYPICAL SHEAR WALL
ONE STORY WALL SECTION STANDARD PLATFORM FRAMING
Residential Design
SDS-CAD
Specialized Design Systems

House #5: 290 ft^2

Residential Design
SDS-CAD
Specialized Design Systems
Custom Cabin Design
(AREA 290 SQUARE FEET)
By Specialized Design
Systems
To the best of my knowledge these plans are drawn to comply with owner's and/ or builder's specifications and any changes made on them after prints are made will be done at the owner's and / or builder's expense and responsibility. The contractor shall verify all dimensions and enclosed drawing. SDSCAD is not liable for errors once construction has begun. While every effort has been made in the preparation of this plan to avoid mistakes, the maker can not guarantee against human error. The contractor of the job must check all dimensions and other details prior to construction and be solely responsible thereafter. All calculations and member sizing should be verified for your building by a certified building official.
Page 1 Cover Page
Page 2 Main Floor Plan
Page 3 Foundation Plan
Page 4 Elevation Plan
Page 5 Floor and Roof Framing Plan
Page 6 Whole House Framing Section
Page 7 Cabinet & Stair Details
Page 8 Electrical Plans
Page 9 Second Floor Plan and ELectrical
Page 10 Details

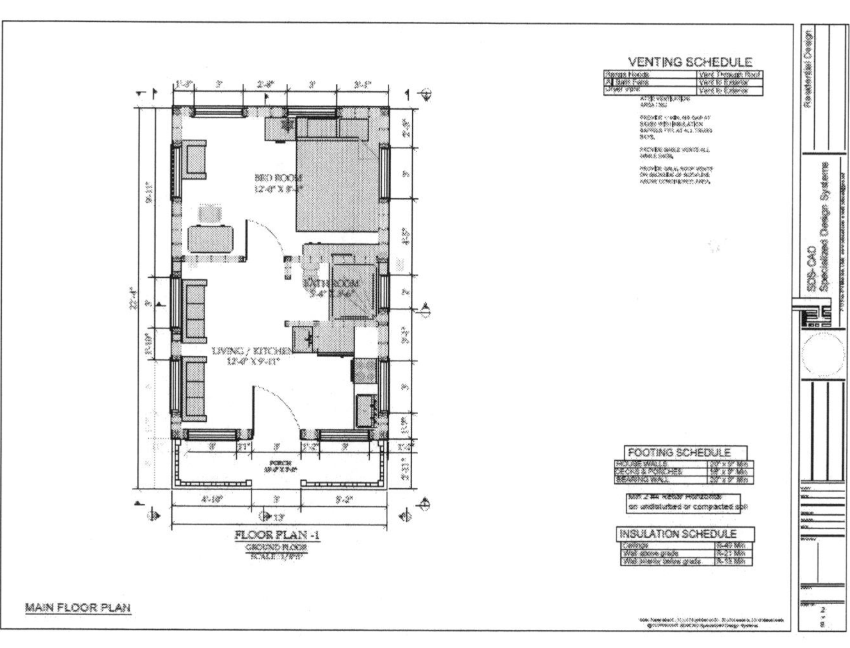

Residential Design
SDS-CAD
Specialized Design Systems
VENTING SCHEDULE
BED ROOM
12'-0" X 8'-1"
BATHROOM
LIVING / KITCHEN
12'-0" X 9'-11"
22'-4"
13'
FLOOR PLAN -1
GROUND FLOOR
FOOTING SCHEDULE
on undisturbed or compacted soil
INSULATION SCHEDULE
MAIN FLOOR PLAN

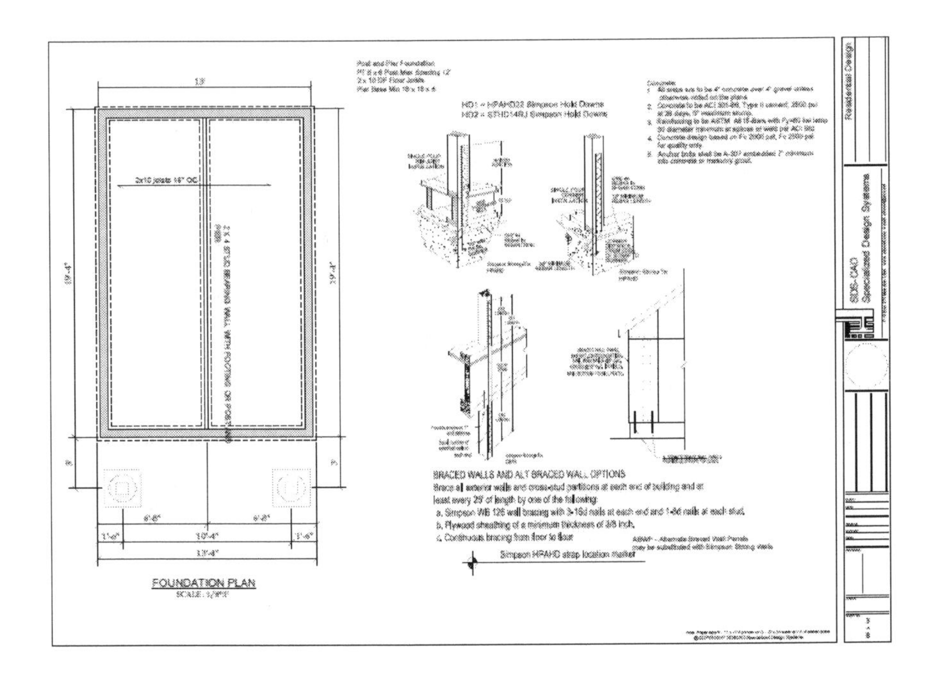

FOUNDATION PLAN
BRACED WALLS AND ALT BRACED WALL OPTIONS
Brace all exterior walls and cross-stud partitions at each end of building and at least every 25' of length by one of the following:
a. Simpson WB 126 wall bracing with 3-16d nails at each end and 1-8d nails at each stud.
b. Plywood sheathing of a minimum thickness of 3/8 inch.
c. Continuous bracing from floor to floor
Simpson HPAHD strap location marker
SDS-CAD
Specialized Design Systems
Residential Design

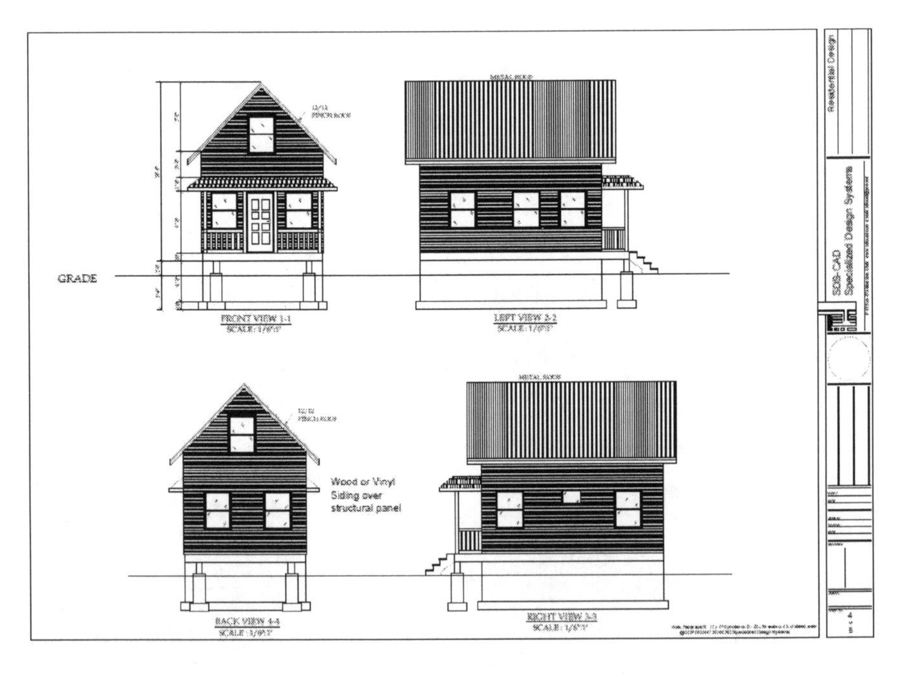

GRADE
FRONT VIEW 1-1
LEFT VIEW 2-2
Wood or Vinyl
Siding over
structural panel
BACK VIEW 4-4
RIGHT VIEW 3-3
SDS-CAD
Specialized Design Systems
Residential Design

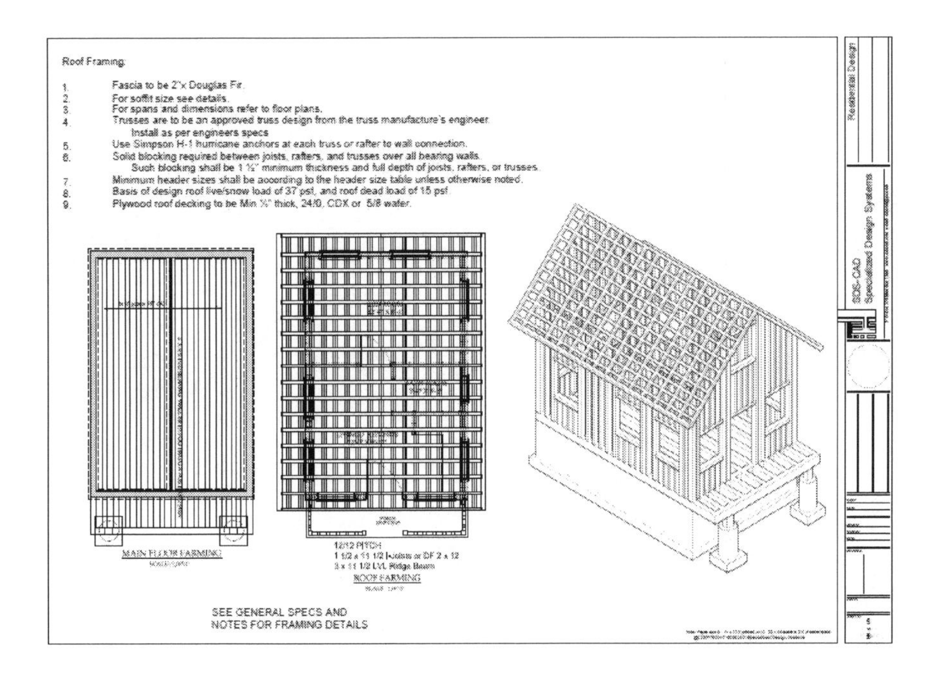

Roof Framing:
1. Fascia to be 2"x Douglas Fir.
2. For soffit size see details.
3. For spans and dimensions refer to floor plans.
4. Trusses are to be an approved truss design from the truss manufacture's engineer.
Install as per engineers specs
5. Use Simpson H-1 hurricane anchors at each truss or rafter to wall connection.
6. Solid blocking required between joists, rafters, and trusses over all bearing walls.
Such blocking shall be 1 ½" minimum thickness and full depth of joists, rafters, or trusses.
7. Minimum header sizes shall be according to the header size table unless otherwise noted.
8. Basis of design roof live/snow load of 37 psf, and roof dead load of 15 psf.
9. Plywood roof decking to be Min ½" thick, 24/0, CDX or 5/8 wafer.
MAIN FLOOR FARMING
12/12 PITCH
1 1/2 x 11 1/2 I-Joists or DF 2 x 12
3 x 11 1/2 LVL Ridge Beam
ROOF FARMING
SEE GENERAL SPECS AND
NOTES FOR FRAMING DETAILS
Residential Design
SDS-CAD
Specialized Design Systems

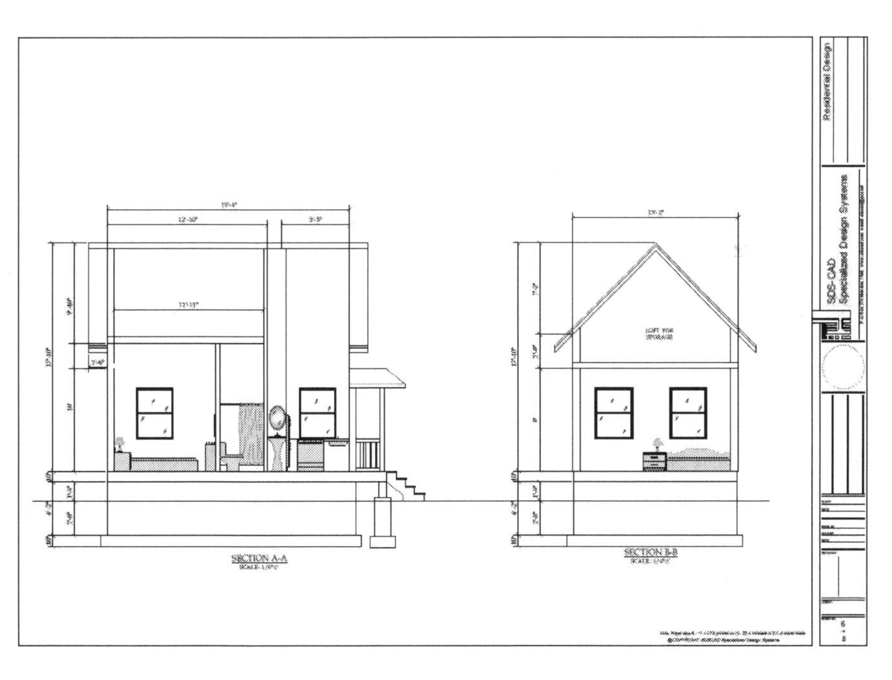

SECTION A-A
SECTION B-B
Residential Design
SDS-CAD
Specialized Design Systems

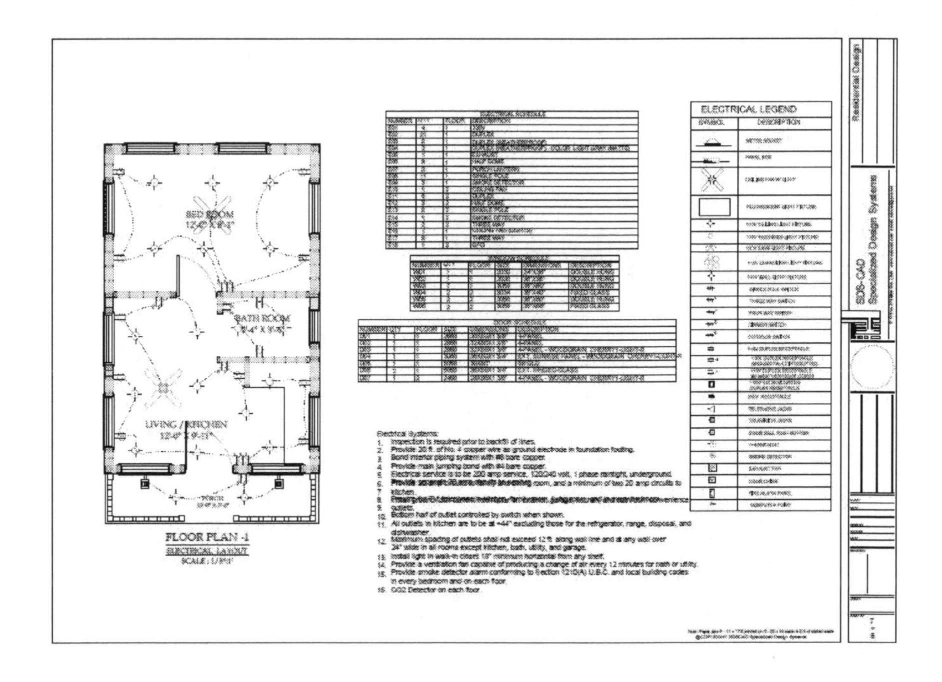
BED ROOM
BATH ROOM
LIVING / KITCHEN
FLOOR PLAN -1
ELECTRICAL LEGEND
Residential Design
SDS-CAD
Specialized Design Systems

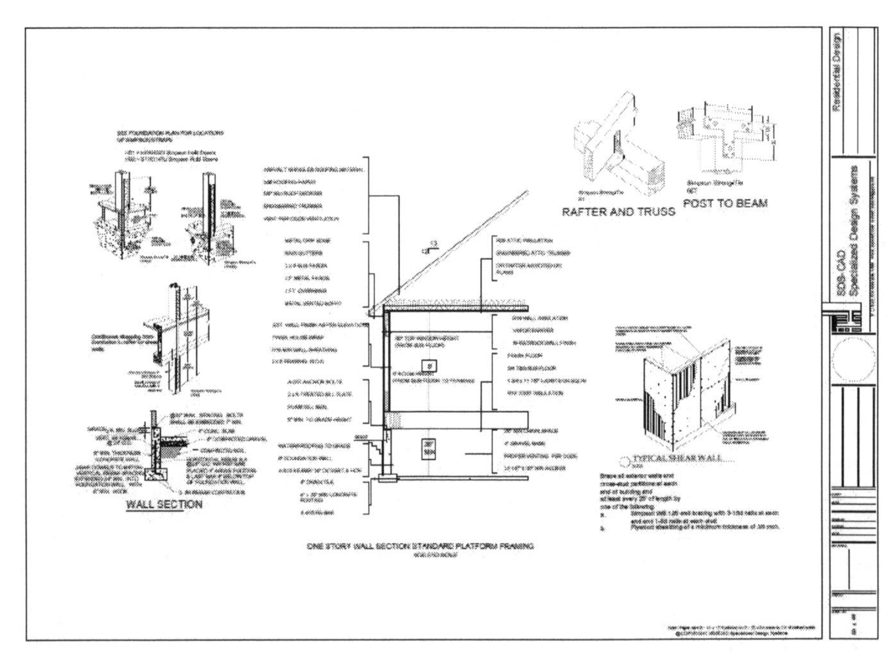
RAFTER AND TRUSS
POST TO BEAM
WALL SECTION
TYPICAL SHEAR WALL
ONE STORY WALL SECTION STANDARD PLATFORM FRAMING
Residential Design
SDS-CAD
Specialized Design Systems

House #6: 325 ft^2

Custom Cabin Design
(AREA 325 SQUIRE FEET)
By Specialized Design Systems
To the best of my knowledge these plans are drawn to comply with owner's and/ or builder's specifications and any changes made on them after prints are made will be done at the owner's and / or builder's expence and responsibility. The contractor shall verify all dimensions and enclosed drawing. SDSCAD is not liable for errors once construction has begun. While every effort has been made in the preparation of this plan to avoid mistakes, the maker can not guarantee against human error. The contractor of the job must check all dimensions and other details prior to construction and be solely responsible thereafter. All calculations and member sizing should be verified for your building by a certified building official.
Page 1 Cover Page
Page 2 Main Floor Plan
Page 3 Foundation Plan
Page 4 Elevation Plan
Page 5 Floor and Roof Framing Plan
Page 6 Whole House Framing Section
Page 7 Cabinet & Stair Details
Page 8 Electrics Plans
Page 9 Second Floor Plan and Electrical
Page 10 Details
Residential Design
SDS-CAD
Specialized Design Systems

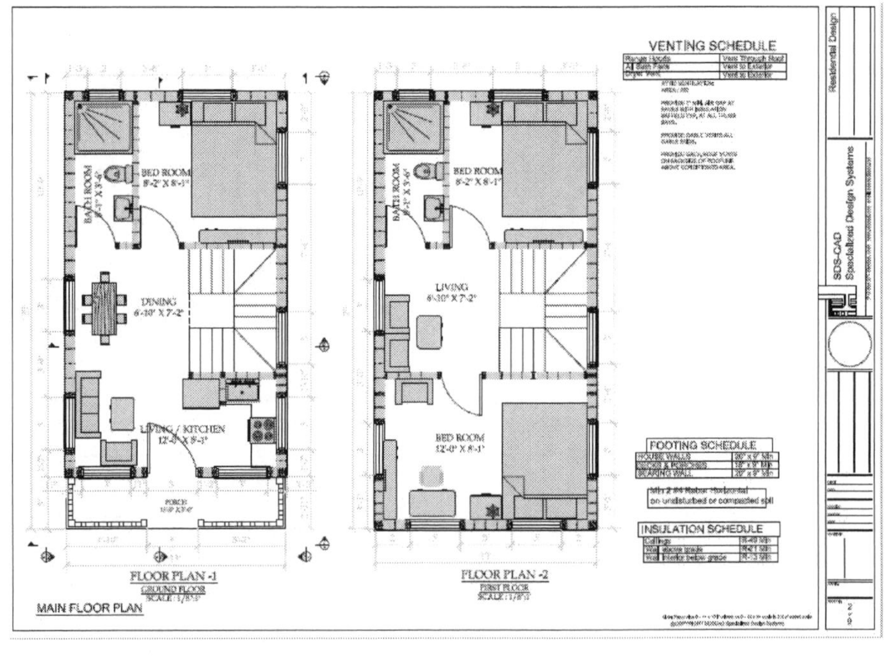
VENTING SCHEDULE
Range Hoods
Vent Through Roof
All Bath Fans
Vent to Exterior
Dryer Vent
Vent to Exterior
BATH ROOM
BED ROOM
8'-2" X 8'-1"
DINING
6'-10" X 7'-2"
LIVING / KITCHEN
LIVING
6'-10" X 7'-2"
BED ROOM
12'-0" X 8'-1"
FOOTING SCHEDULE
INSULATION SCHEDULE
FLOOR PLAN -1
GROUND FLOOR
SCALE : 1/8"=1'
FLOOR PLAN -2
FIRST FLOOR
SCALE : 1/8"=1'
MAIN FLOOR PLAN
Residential Design
SDS-CAD
Specialized Design Systems

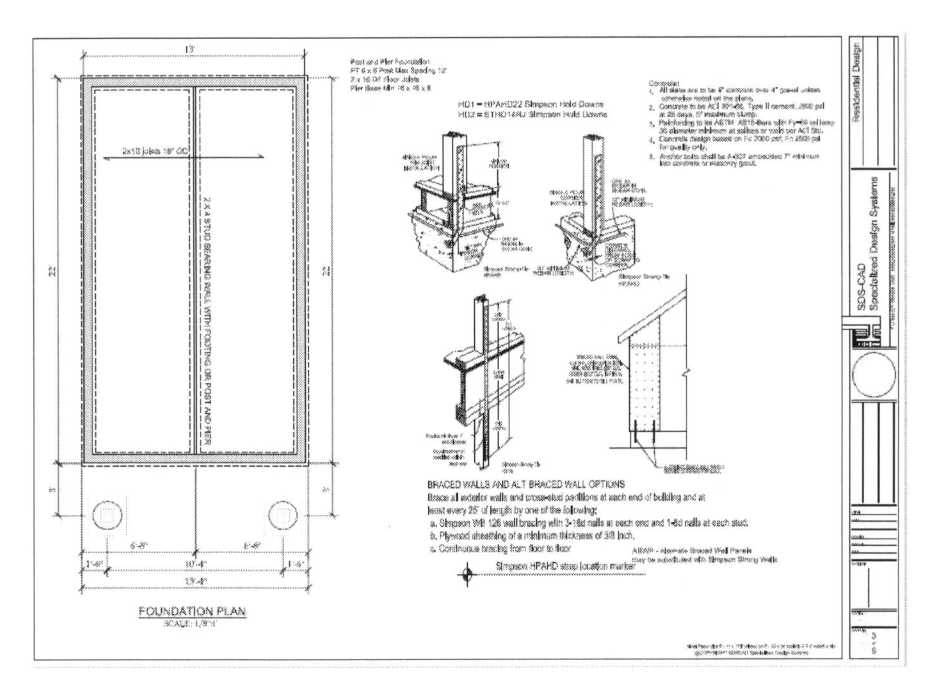
Post and Pier Foundation
HD1 = HPAHD22 Simpson Hold Downs
BRACED WALLS AND ALT BRACED WALL OPTIONS
Brace all exterior walls and cross-stud partitions at each end of building and at least every 25' of length by one of the following:
a. Simpson WB 126 wall bracing with 3-16d nails at each end and 1-8d nails at each stud.
b. Plywood sheathing of a minimum thickness of 3/8 inch.
c. Continuous bracing from floor to floor
Simpson HPAHD strap location marker
FOUNDATION PLAN
Residential Design
SDS-CAD
Specialized Design Systems

GRADE
FRONT VIEW 1-1
LEFT VIEW 2-2
METAL ROOF
Wood or Vinyl
Siding over
structural panel
BACK VIEW 4-4
RIGHT VIEW 3-3
Residential Design
SDS-CAD
Specialized Design Systems

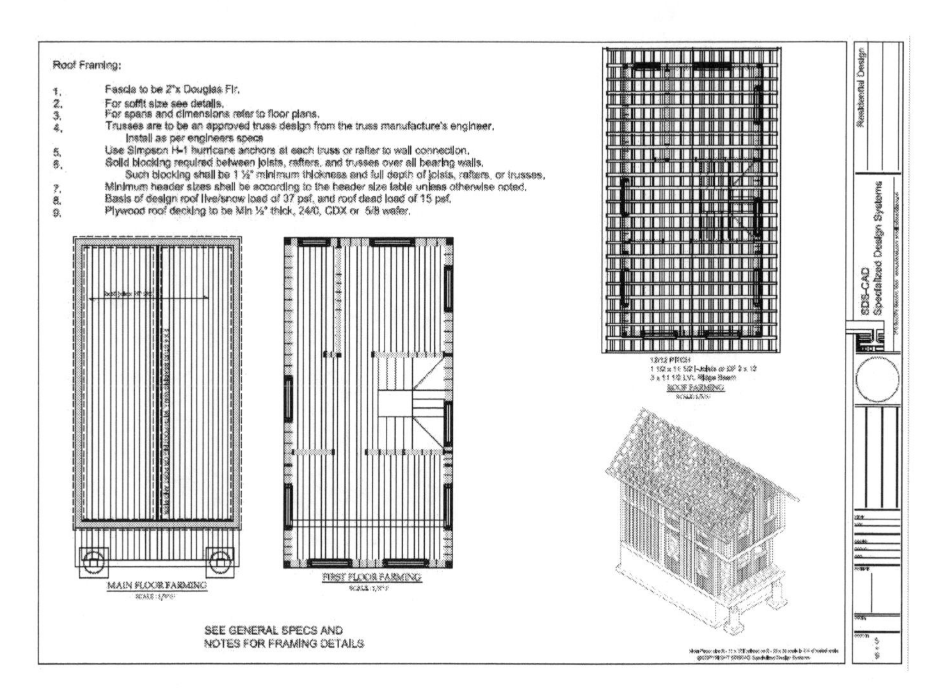

Roof Framing:
1. Fascia to be 2"x Douglas Fir.
2. For soffit size see details.
3. For spans and dimensions refer to floor plans.
4. Trusses are to be an approved truss design from the truss manufacture's engineer. Install as per engineers specs
5. Use Simpson H-1 hurricane anchors at each truss or rafter to wall connection.
6. Solid blocking required between joists, rafters, and trusses over all bearing walls. Such blocking shall be 1 ½" minimum thickness and full depth of joists, rafters, or trusses.
7. Minimum header sizes shall be according to the header size table unless otherwise noted.
8. Basis of design roof live/snow load of 37 psf, and roof dead load of 15 psf.
9. Plywood roof decking to be Min ½" thick, 24/0, CDX or 5/8 wafer.
MAIN FLOOR FARMING
FIRST FLOOR FARMING
ROOF FARMING
SEE GENERAL SPECS AND
NOTES FOR FRAMING DETAILS
Residential Design
SDS-CAD
Specialized Design Systems

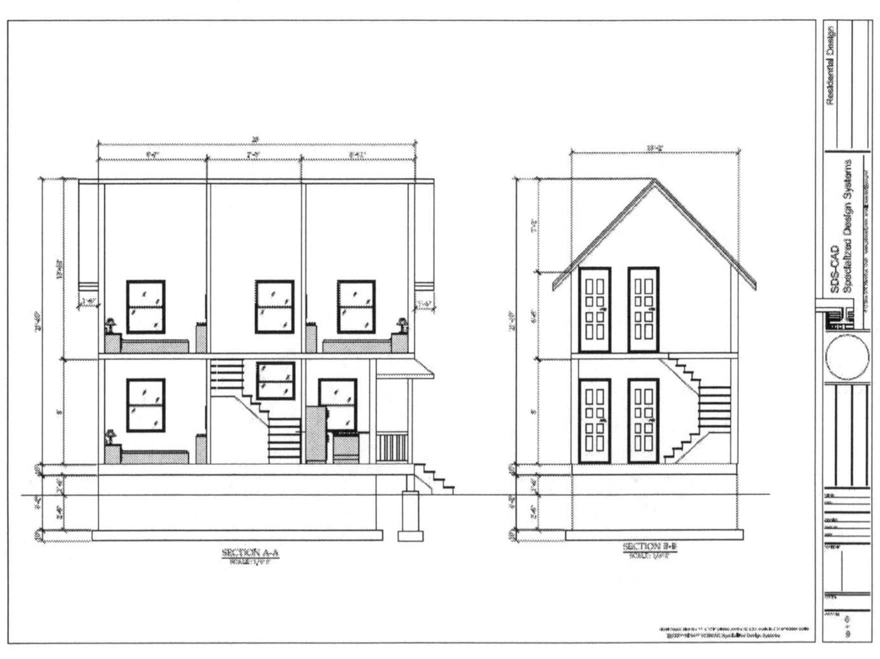

SECTION A-A
SECTION B-B
Residential Design
SDS-CAD
Specialized Design Systems

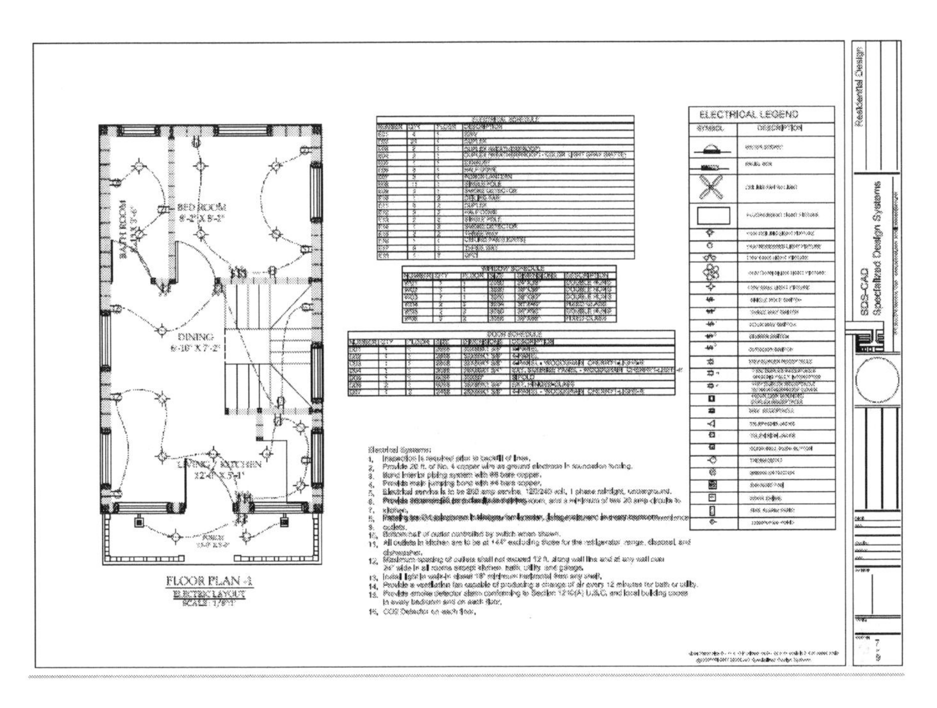
FLOOR PLAN -1
ELECTRICAL LEGEND
SDS-CAD
Specialized Design Systems
Residential Design

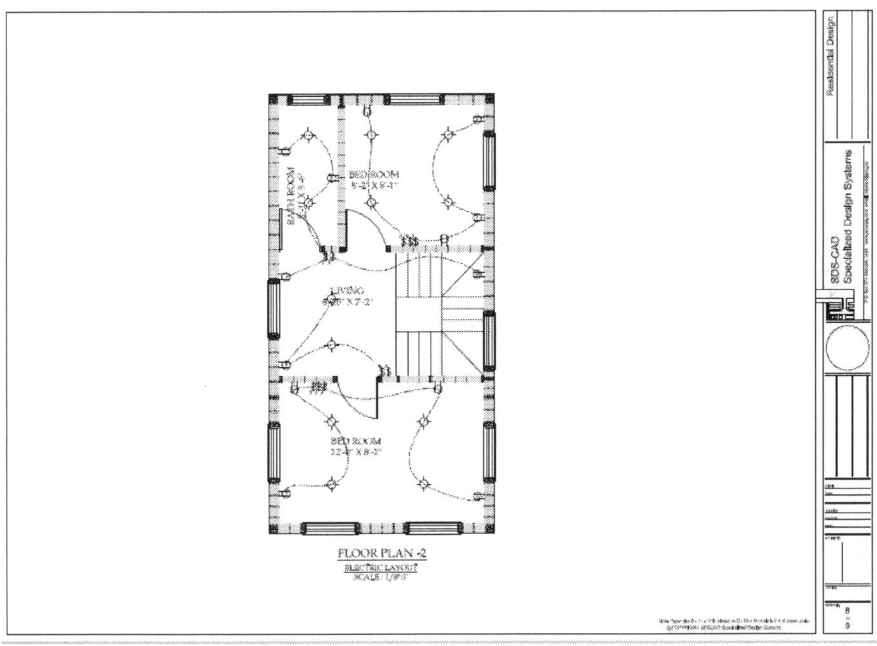
FLOOR PLAN -2
BED ROOM
LIVING
SDS-CAD
Specialized Design Systems
Residential Design

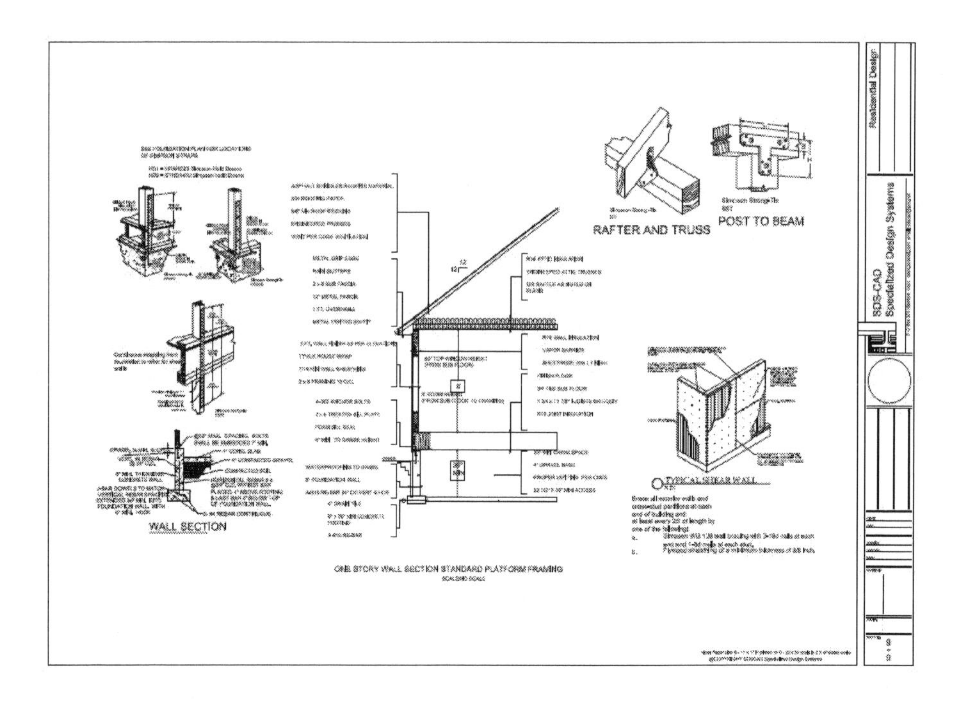
RAFTER AND TRUSS
POST TO BEAM
WALL SECTION
ONE STORY WALL SECTION STANDARD PLATFORM FRAMING
TYPICAL SHEAR WALL
Residential Design
SDS-CAD
Specialized Design Systems

House #7: 435 ft^2

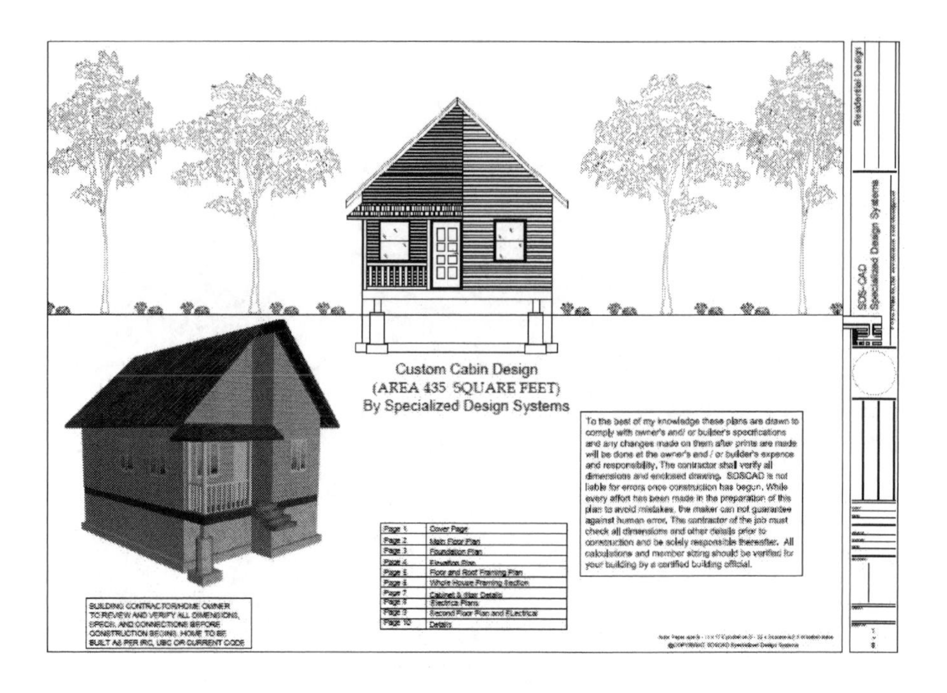

Custom Cabin Design
(AREA 435 SQUARE FEET)
By Specialized Design Systems
To the best of my knowledge these plans are drawn to comply with owner's and/ or builder's specifications and any changes made on them after prints are made will be done at the owner's and / or builder's expence and responsibility. The contractor shall verify all dimensions and enclosed drawing. SDSCAD is not liable for errors once construction has begun. While every effort has been made in the preparation of this plan to avoid mistakes, the maker can not guarantee against human error. The contractor of the job must check all dimensions and other details prior to construction and be solely responsible thereafter. All calculations and member sizing should be verified for your building by a certified building official.
Page 1 Cover Page
Page 2 Main Floor Plan
Page 3 Foundation Plan
Page 4 Elevation Plan
Page 5 Floor and Roof Framing Plan
Page 6 Whole House Framing Section
Page 7 Cabinet & Stair Details
Page 8 Electrical Plans
Page 9 Second Floor Plan and ELectrical
Page 10 Details
BUILDING CONTRACTOR/HOME OWNER TO REVIEW AND VERIFY ALL DIMENSIONS, SPECS. AND CONNECTIONS BEFORE CONSTRUCTION BEGINS. HOME TO BE BUILT AS PER IRC, UBC OR CURRENT CODE
Residential Design
SDS-CAD
Specialized Design Systems

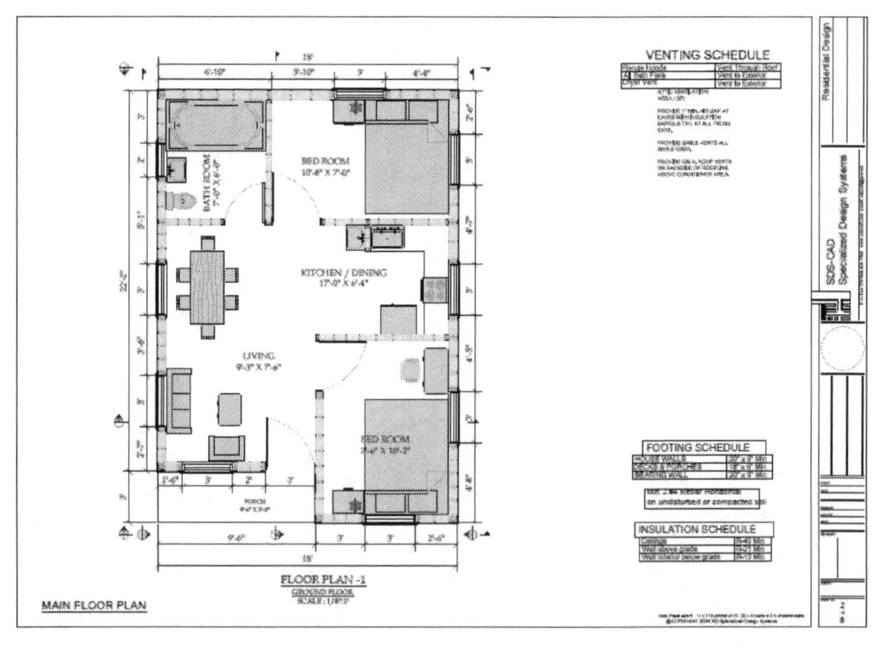

VENTING SCHEDULE
FOOTING SCHEDULE
INSULATION SCHEDULE
BED ROOM
10'-8" X 7'-0"
BATH ROOM
7'-0" X 6'-0"
KITCHEN / DINING
17'-0" X 6'-4"
LIVING
9'-3" X 7'-6"
BED ROOM
7'-6" X 10'-2"
PORCH
FLOOR PLAN -1
GROUND FLOOR
SCALE : 1/8"=1'
MAIN FLOOR PLAN
Residential Design
SDS-CAD
Specialized Design Systems

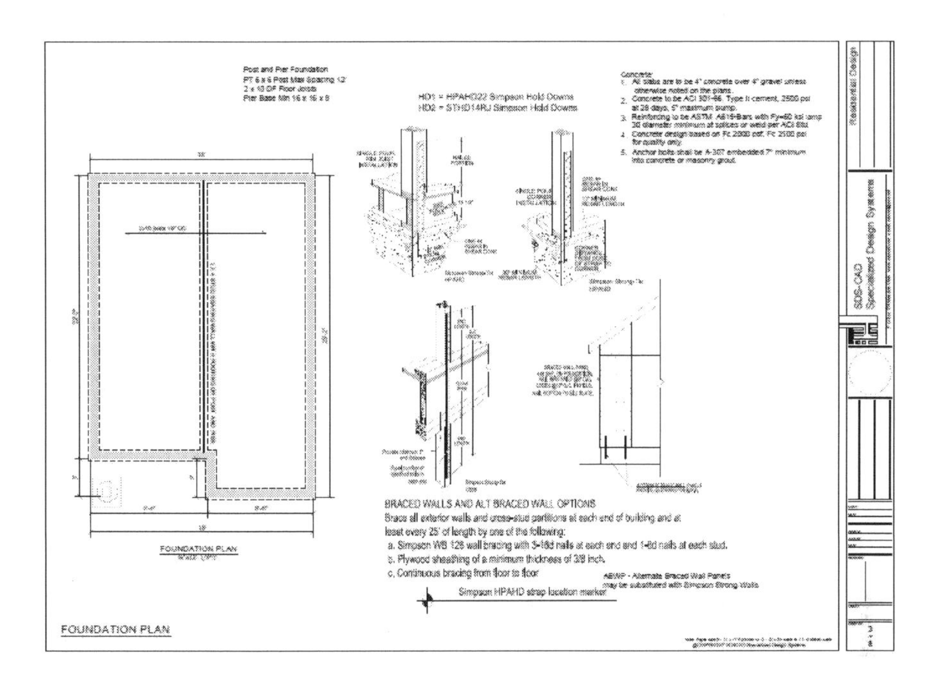
Post and Pier Foundation
PT 6 x 6 Post Max Spacing 12'
2 x 10 DF Floor Joists
Pier Base Min 16 x 16 x 8
HD1 = HPAHD22 Simpson Hold Downs
HD2 = STHD14RJ Simpson Hold Downs
FOUNDATION PLAN
BRACED WALLS AND ALT BRACED WALL OPTIONS
Brace all exterior walls and cross-stud partitions at each end of building and at
least every 25' of length by one of the following:
a. Simpson WB 126 wall bracing with 3-16d nails at each end and 1-8d nails at each stud.
b. Plywood sheathing of a minimum thickness of 3/8 inch.
c. Continuous bracing from floor to floor
ABWP - Alternate Braced Wall Panels
may be substituted with Simpson Strong Walls
Simpson HPAHD strap location marker
FOUNDATION PLAN
Residential Design
SDS-CAD
Specialized Design Systems

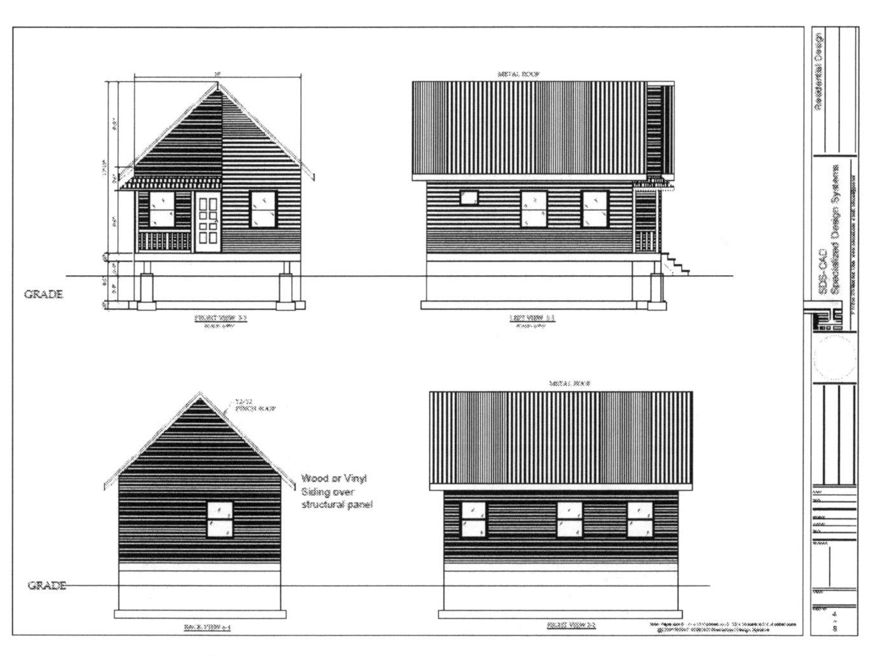
METAL ROOF
GRADE
METAL ROOF
Wood or Vinyl
Siding over
structural panel
GRADE
Residential Design
SDS-CAD
Specialized Design Systems

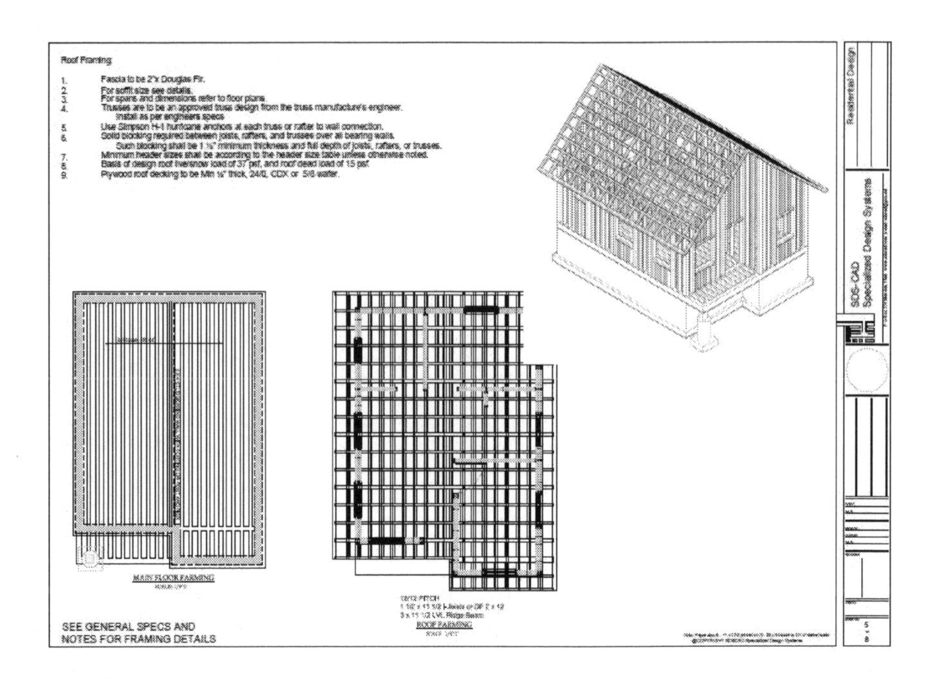
Roof Framing
1. Fascia to be 2"x Douglas Fir.
2. For soffit size see details.
3. For spans and dimensions refer to floor plans.
4. Trusses are to be an approved truss design from the truss manufacture's engineer.
Install as per engineers specs
5. Use Simpson H-1 hurricane anchors at each truss or rafter to wall connection.
6. Solid blocking required between joists, rafters, and trusses over all bearing walls.
Such blocking shall be 1 ½" minimum thickness and full depth of joists, rafters, or trusses.
7. Minimum header sizes shall be according to the header size table unless otherwise noted.
8. Basis of design roof live/snow load of 37 psf, and roof dead load of 15 psf.
9. Plywood roof decking to be Min ½" thick, 24/0, CDX or 5/8 wafer.
MAIN FLOOR FARMING
12/12 PITCH
3 x 11 1/2 LVL Ridge Beam
ROOF FARMING
SEE GENERAL SPECS AND
NOTES FOR FRAMING DETAILS
Residential Design
SDS-CAD
Specialized Design Systems

SECTION A-A
SECTION B-B
SECTION DETAILS
Residential Design
SDS-CAD
Specialized Design Systems

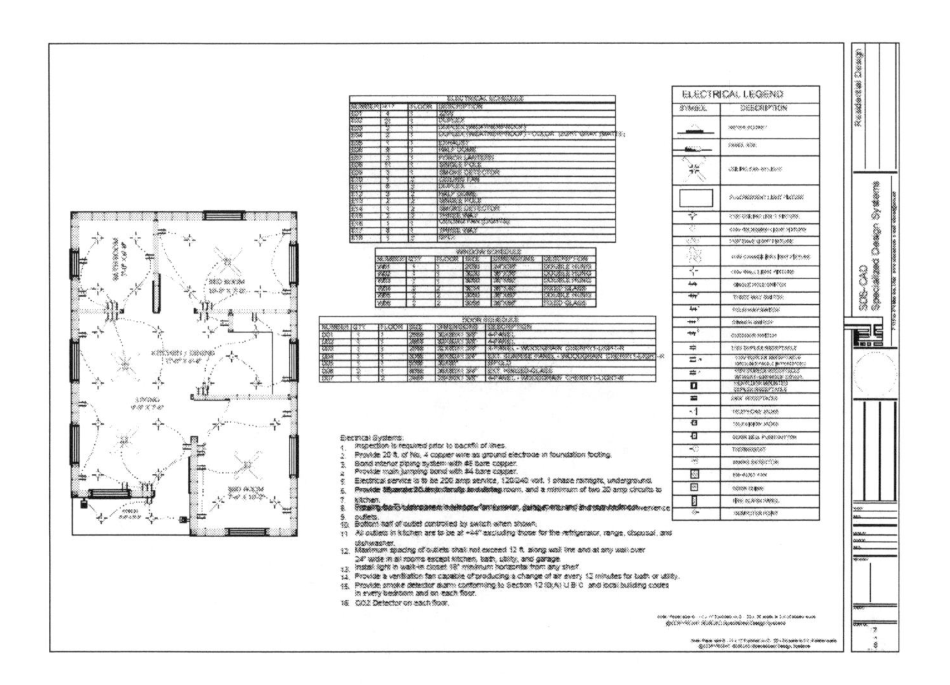

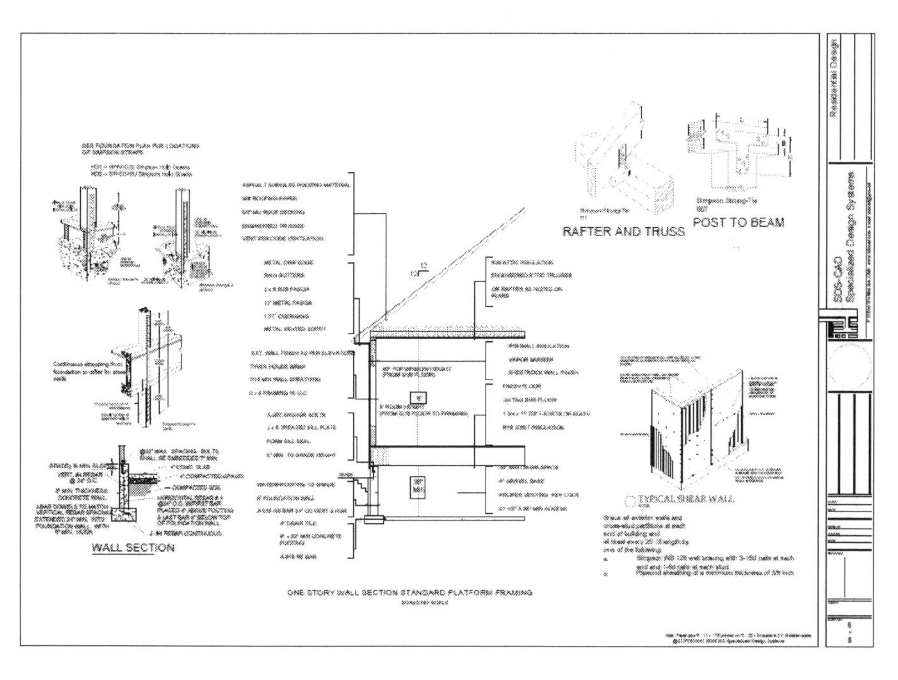
RAFTER AND TRUSS
POST TO BEAM
WALL SECTION
TYPICAL SHEAR WALL
ONE STORY WALL SECTION STANDARD PLATFORM FRAMING
Residential Design
SDS-CAD
Specialized Design Systems

House #8: 435 ft^2

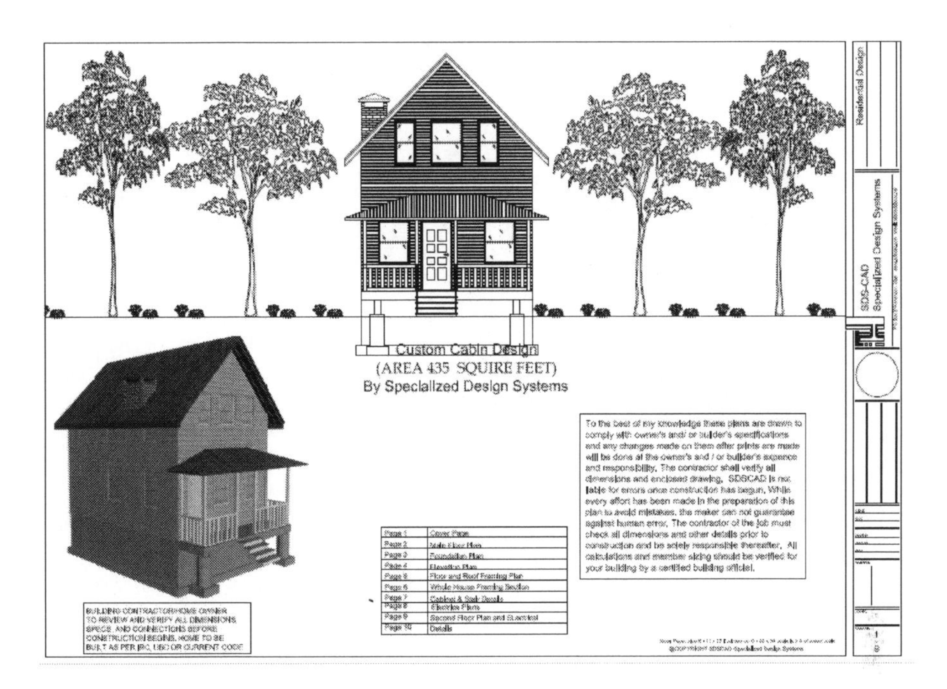
Custom Cabin Design
(AREA 435 SQUIRE FEET)
By Specialized Design Systems
To the best of my knowledge these plans are drawn to comply with owner's and/ or builder's specifications and any changes made on them after prints are made will be done at the owner's and / or builder's expence and responsibility. The contractor shall verify all dimensions and enclosed drawing. SDSCAD is not liable for errors once construction has begun. While every effort has been made in the preparation of this plan to avoid mistakes, the maker can not guarantee against human error. The contractor of the job must check all dimensions and other details prior to construction and be solely responsible thereafter. All calculations and member sizing should be verified for your building by a certified building official.
Page 1 Cover Page
Page 2 Main Floor Plan
Page 3 Foundation Plan
Page 4 Elevation Plan
Page 5 Floor and Roof Framing Plan
Page 6 Whole House Framing Section
Page 7 Cabinet & Stair Details
Page 8 Electrical Plans
Page 9 Second Floor Plan and Electrical
Page 10 Details
BUILDING CONTRACTOR/HOME OWNER TO REVIEW AND VERIFY ALL DIMENSIONS SPECS. AND CONNECTIONS BEFORE CONSTRUCTION BEGINS. HOME TO BE BUILT AS PER IRC, UBC OR CURRENT CODE
Residential Design
SDS-CAD
Specialized Design Systems

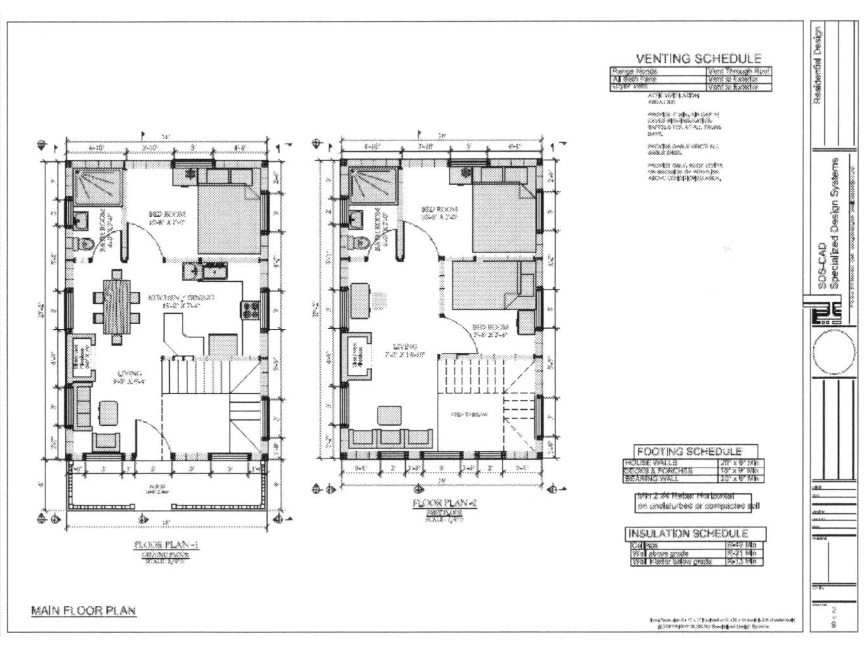
VENTING SCHEDULE
FOOTING SCHEDULE
INSULATION SCHEDULE
FLOOR PLAN-1
FLOOR PLAN-2
MAIN FLOOR PLAN
Residential Design
SDS-CAD
Specialized Design Systems

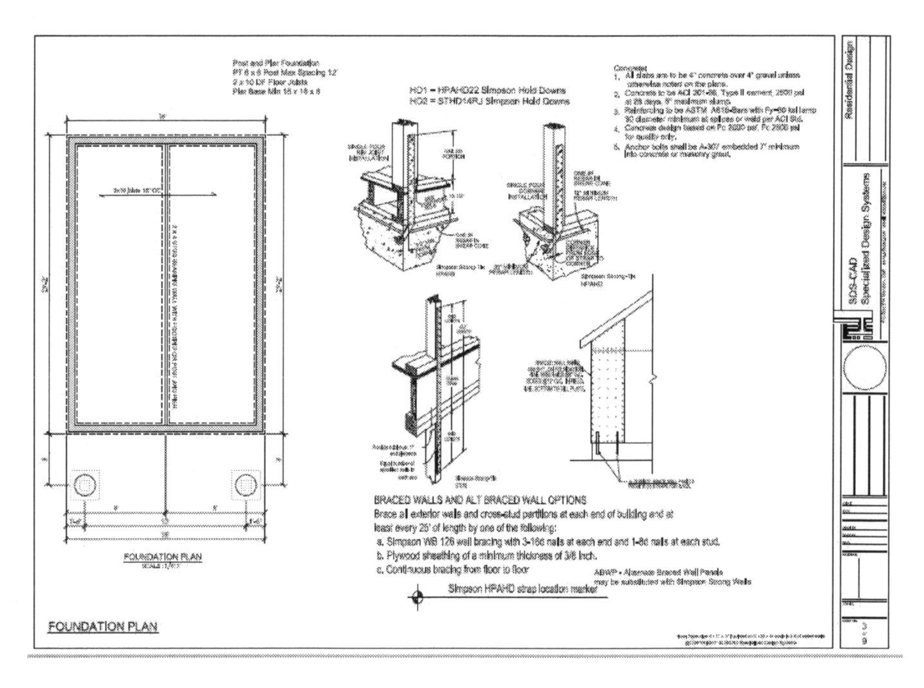
Post and Pier Foundation
PT 6 x 6 Post Max Spacing 12'
2 x 10 DF Floor Joists
HO1 = HPAHD22 Simpson Hold Downs
HO2 = STHD14RJ Simpson Hold Downs
BRACED WALLS AND ALT BRACED WALL OPTIONS
Brace all exterior walls and cross-stud partitions at each end of building and at least every 25' of length by one of the following:
a. Simpson WB 126 wall bracing with 3-16d nails at each end and 1-8d nails at each stud.
b. Plywood sheathing of a minimum thickness of 3/8 inch.
c. Continuous bracing from floor to floor
Simpson HPAHD strap location marker
FOUNDATION PLAN
FOUNDATION PLAN
Residential Design
SDS-CAD
Specialized Design Systems

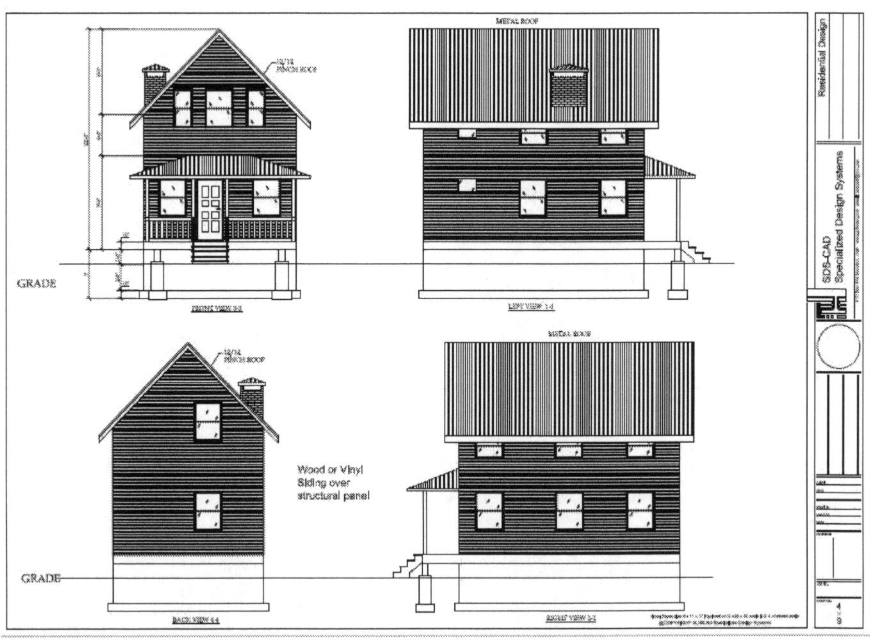
METAL ROOF
GRADE
METAL ROOF
Wood or Vinyl
Siding over
structural panel
GRADE
Residential Design
SDS-CAD
Specialized Design Systems

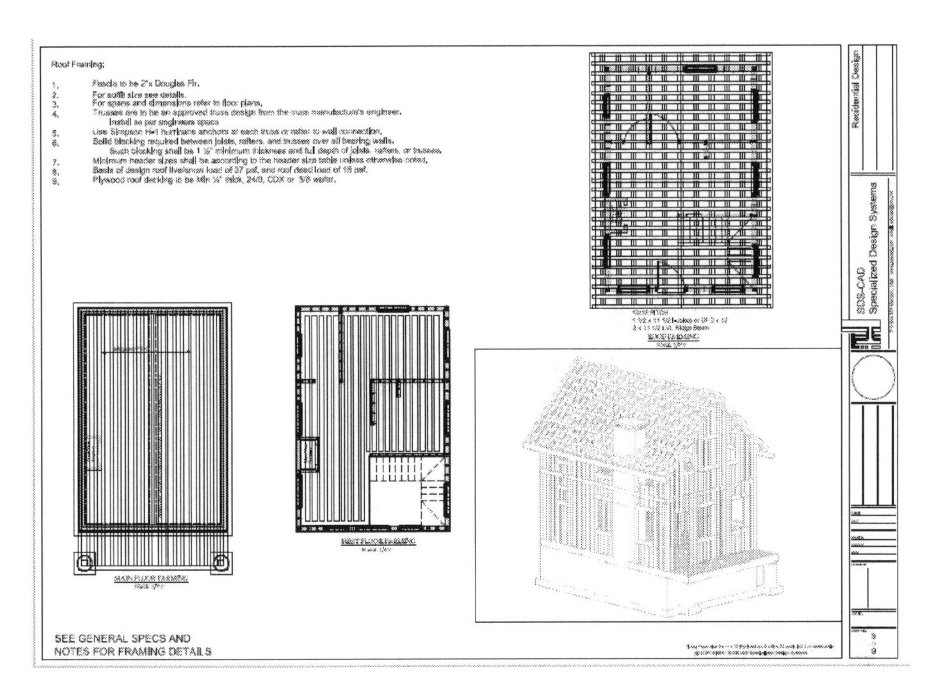

Roof Framing;
1. Fascia to be 2"x Douglas Fir.
2. For soffit size see details.
3. For spans and dimensions refer to floor plans.
7. Minimum header sizes shall be according to the header size table unless otherwise noted.
Residential Design
SDS-CAD
Specialized Design Systems
SEE GENERAL SPECS AND
NOTES FOR FRAMING DETAILS

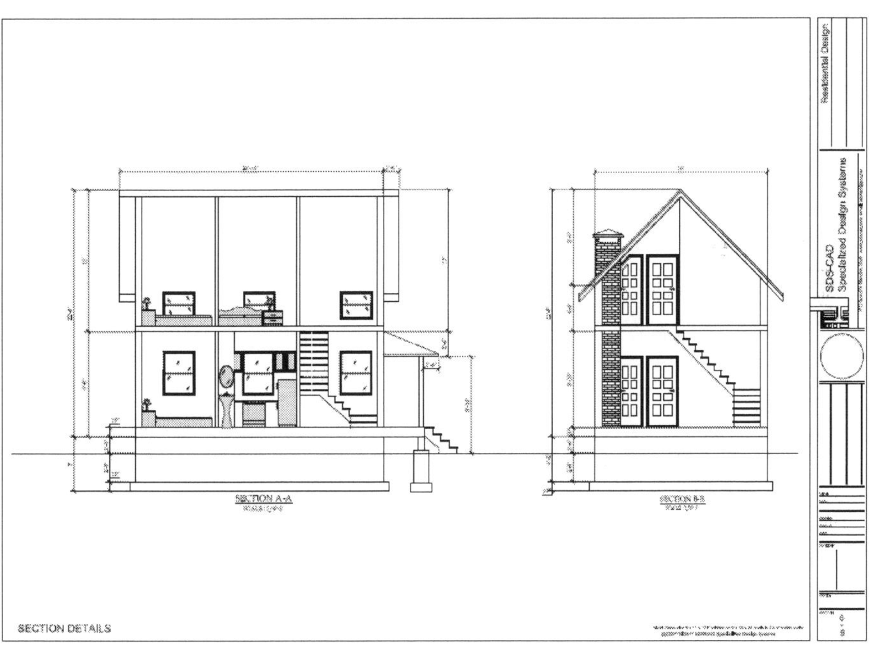

Residential Design
SDS-CAD
Specialized Design Systems
SECTION A-A
SECTION B-B
SECTION DETAILS

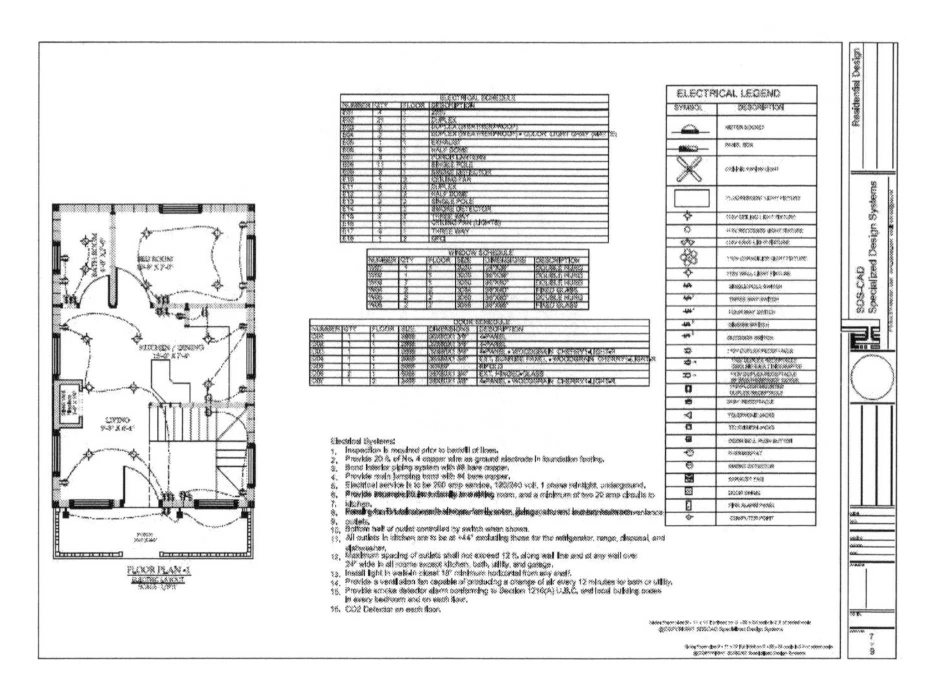
ELECTRICAL LEGEND
SYMBOL
DESCRIPTION
FLOOR PLAN -1
Electrical Systems:
1. Inspection is required prior to backfill of lines.
2. Provide 20 ft. of No. 4 copper wire as ground electrode in foundation footing.
3. Bond interior piping system with #8 bare copper.
4. Provide main jumping bond with #4 bare copper.
5. Electrical service is to be 200 amp service, 120/240 volt, 1 phase raintight, underground.
10. Bottom half of outlet controlled by switch when shown.
11. All outlets in kitchen are to be at +44" excluding those for the refrigerator, range, disposal, and dishwasher.
12. Maximum spacing of outlets shall not exceed 12 ft. along wall line and at any wall over 24" wide in all rooms except kitchen, bath, utility, and garage.
13. Install light in walk-in closet 18" minimum horizontal from any shelf.
14. Provide a ventilation fan capable of producing a change of air every 12 minutes for bath or utility.
15. Provide smoke detector alarm conforming to Section 1210(A) U.B.C. and local building codes in every bedroom and on each floor.
16. CO2 Detector on each floor.
Residential Design
SDS-CAD
Specialized Design Systems

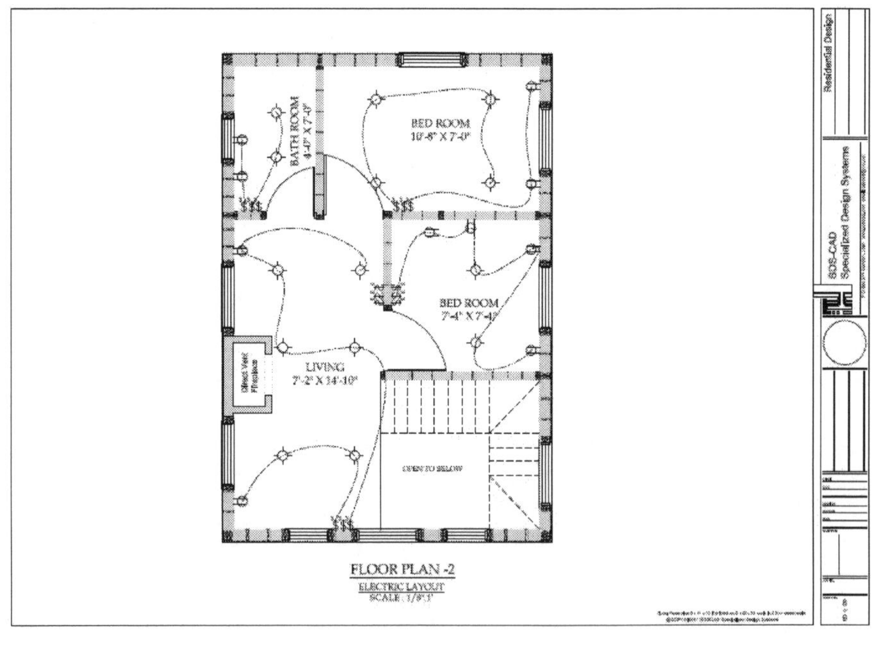
BATH ROOM
4'-0" X 7'-0"
BED ROOM
10'-8" X 7'-0"
BED ROOM
7'-4" X 7'-4"
LIVING
7'-2" X 14'-10"
Direct Vent Fireplace
OPEN TO BELOW
FLOOR PLAN -2
ELECTRIC LAYOUT
SCALE : 1/8"-1'
Residential Design
SDS-CAD
Specialized Design Systems

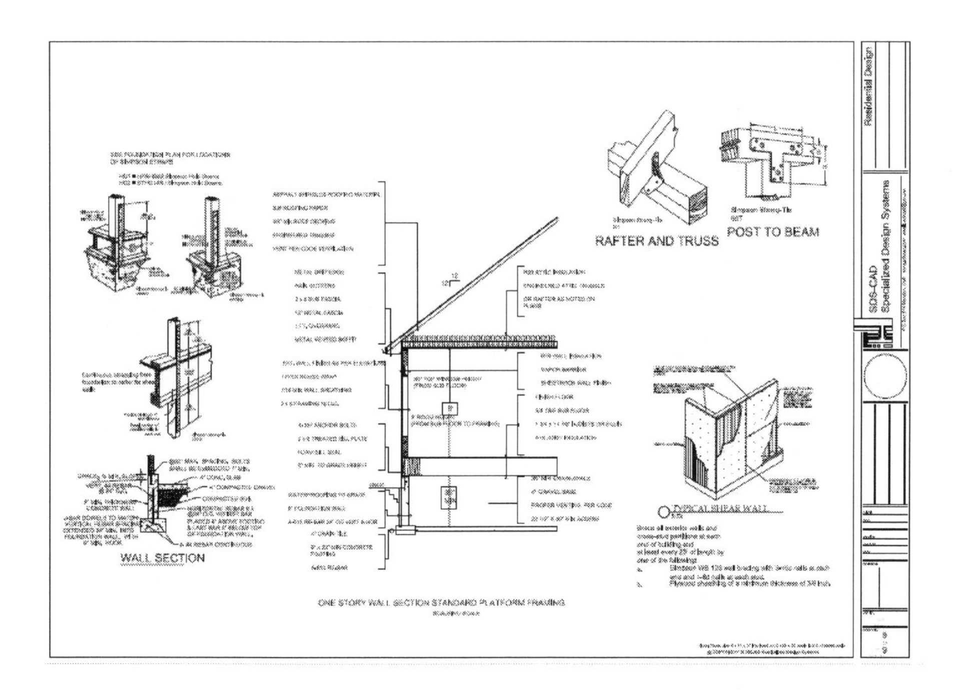
RAFTER AND TRUSS
POST TO BEAM
WALL SECTION
ONE STORY WALL SECTION STANDARD PLATFORM FRAMING
Residential Design
SDS-CAD
Specialized Design Systems

House #9: 520 ft^2

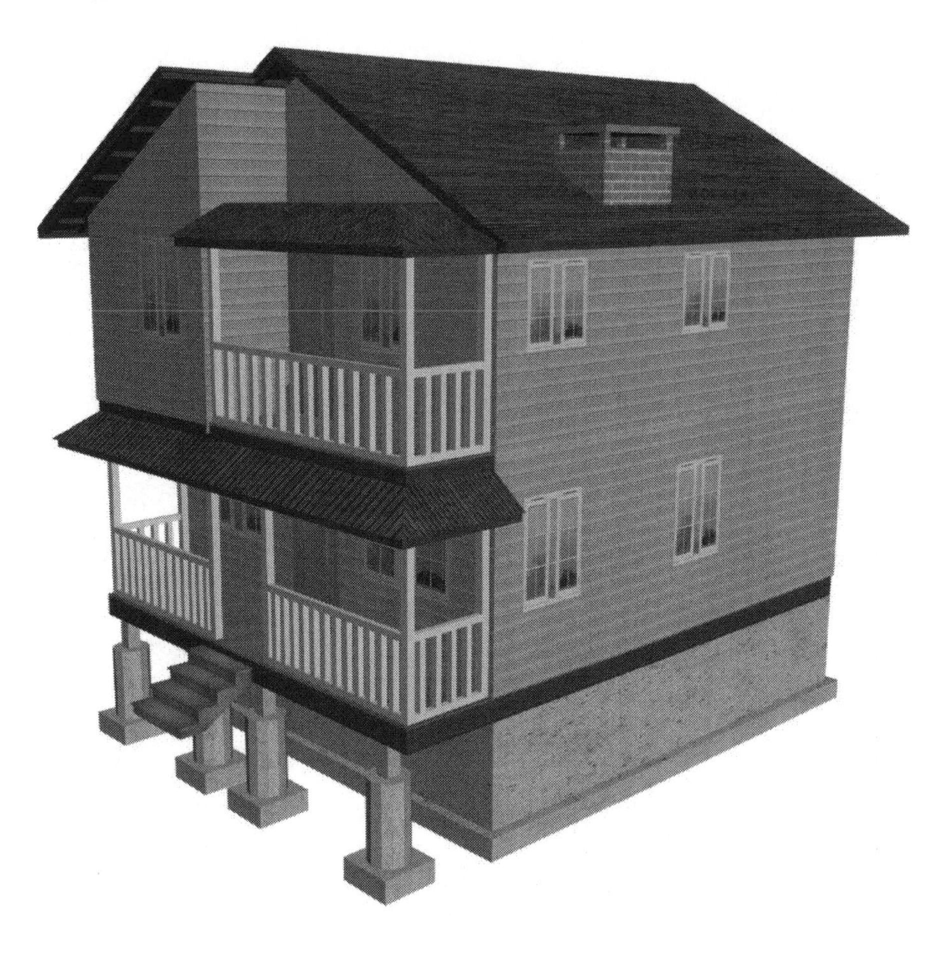

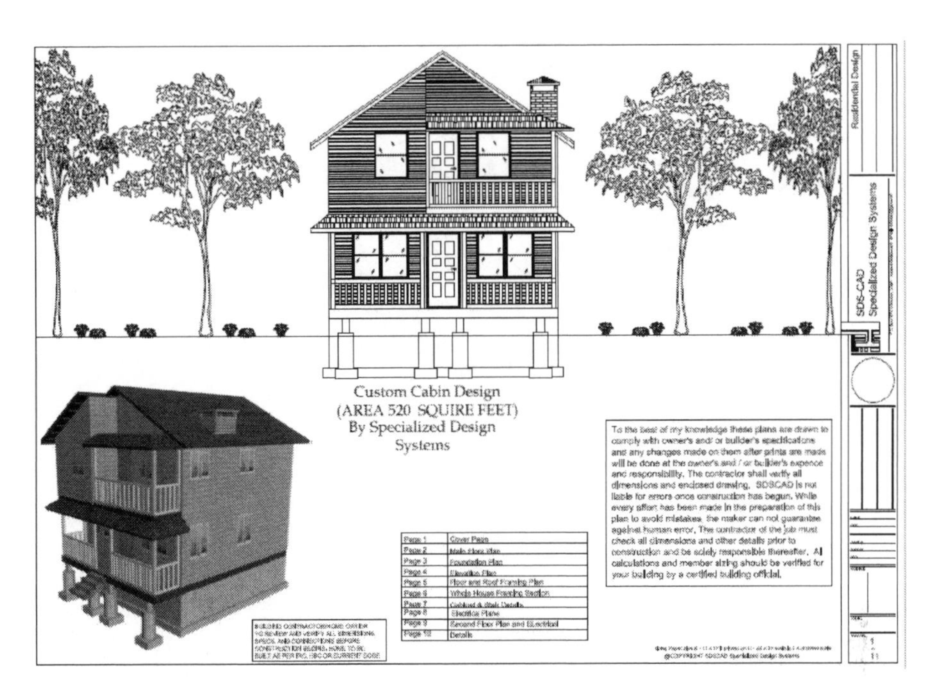
Residential Design
SDS-CAD
Specialized Design Systems
Custom Cabin Design
(AREA 520 SQUIRE FEET)
By Specialized Design
Systems
To the best of my knowledge these plans are drawn to comply with owner's and/ or builder's specifications and any changes made on them after prints are made will be done at the owner's and / or builder's expence and responsibility. The contractor shall verify all dimensions and enclosed drawing. SDSCAD is not liable for errors once construction has begun. While every effort has been made in the preparation of this plan to avoid mistakes, the maker can not guarantee against human error. The contractor of the job must check all dimensions and other details prior to construction and be solely responsible thereafter. Al calculations and member sizing should be verified for your building by a certified building official.
Page 1 Cover Page
Page 2 Main Floor Plan
Page 3 Foundation Plan
Page 4 Elevation Plan
Page 5 Floor and Roof Framing Plan
Page 6 Whole House Framing Section
Page 7 Cabinet & Stair Details
Page 8 Electrical Plans
Page 9 Second Floor Plan and Electrical
Page 10 Details

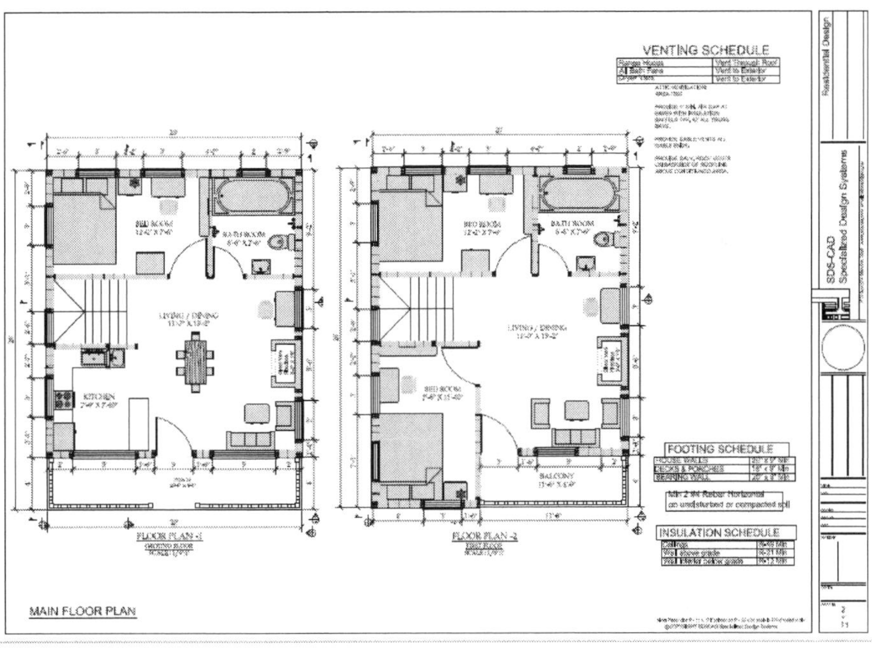
Residential Design
SDS-CAD
Specialized Design Systems
VENTING SCHEDULE
Range Hoods | Vent Through Roof
All Bath Fans | Vent to Exterior
Dryer Vent | Vent to Exterior
BED ROOM
BATH ROOM
LIVING / DINING
KITCHEN
BALCONY
FLOOR PLAN -1
FLOOR PLAN -2
FOOTING SCHEDULE
Min 2 #4 Rebar Horizontal
on undisturbed or compacted soil
INSULATION SCHEDULE
Ceilings
Wall above grade
Wall interior below grade
MAIN FLOOR PLAN

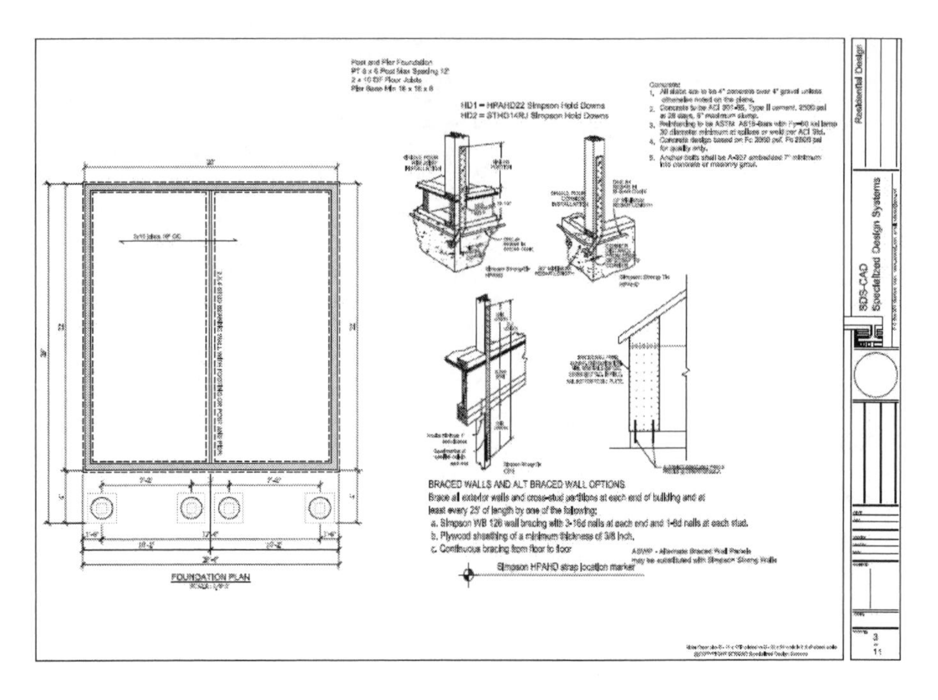
Post and Pier Foundation
HD1 = HPAHD22 Simpson Hold Downs
HD2 = STHD14RJ Simpson Hold Downs
FOUNDATION PLAN
BRACED WALLS AND ALT BRACED WALL OPTIONS
Brace all exterior walls and cross-stud partitions at each end of building and at
least every 25' of length by one of the following:
a. Simpson WB 126 wall bracing with 3-16d nails at each end and 1-8d nails at each stud.
b. Plywood sheathing of a minimum thickness of 3/8 inch.
c. Continuous bracing from floor to floor
Simpson HPAHD strap location marker
Residential Design
SDS-CAD
Specialized Design Systems

METAL ROOF
PINCH ROOF
GRADE
FRONT VIEW 1-1
LEFT VIEW 2-2
Wood or Vinyl
Siding over
structural panel
Residential Design
SDS-CAD
Specialized Design Systems

GRADE
BACK VIEW 4-4
RIGHT VIEW 3-3
Wood or Vinyl
Siding over
structural panel
Residential Design
SDS-CAD
Specialized Design Systems

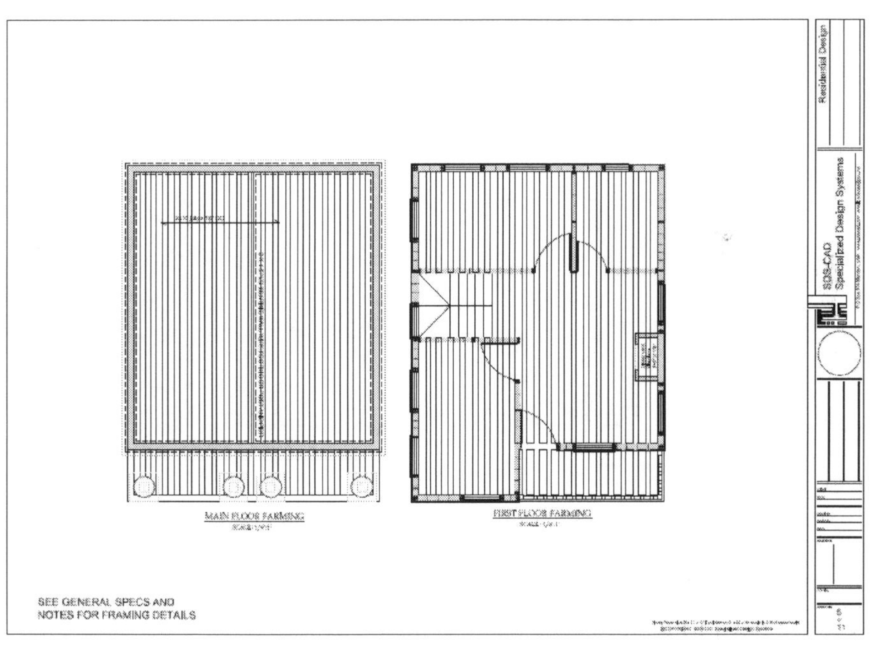
MAIN FLOOR FARMING
FIRST FLOOR FARMING
SEE GENERAL SPECS AND
NOTES FOR FRAMING DETAILS
Residential Design
SDS-CAD
Specialized Design Systems

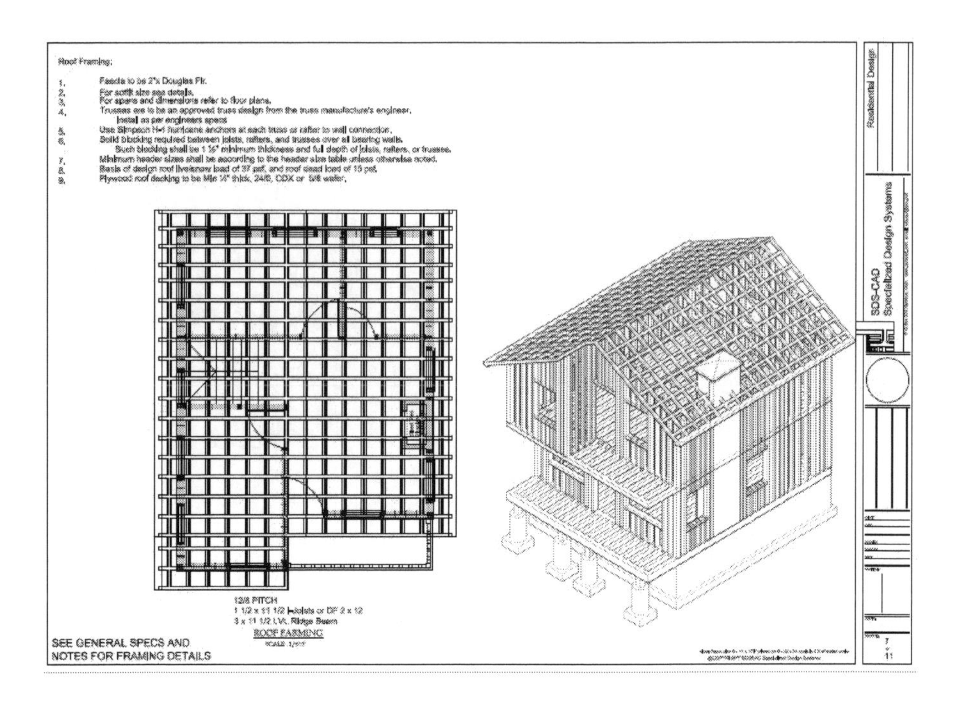
Roof Framing;
1. Fascia to be 2"x Douglas Fir.
2. For soffit size see details.
3. For spans and dimensions refer to floor plans.
4. Trusses are to be an approved truss design from the truss manufacture's engineer. Install as per engineers specs
5. Use Simpson H-1 hurricane anchors at each truss or rafter to wall connection.
6. Solid blocking required between joists, rafters, and trusses over all bearing walls. Such blocking shall be 1 ½" minimum thickness and full depth of joists, rafters, or trusses.
7. Minimum header sizes shall be according to the header size table unless otherwise noted.
8. Basis of design roof live/snow load of 37 psf, and roof dead load of 15 psf.
9. Plywood roof decking to be Min ½" thick, 24/0, CDX or 5/8 wafer.
12/8 PITCH
1 1/2 x 11 1/2 I-Joists or DF 2 x 12
3 x 11 1/2 LVL Ridge Beam
ROOF FARMING
SEE GENERAL SPECS AND
NOTES FOR FRAMING DETAILS
Residential Design
SDS-CAD
Specialized Design Systems

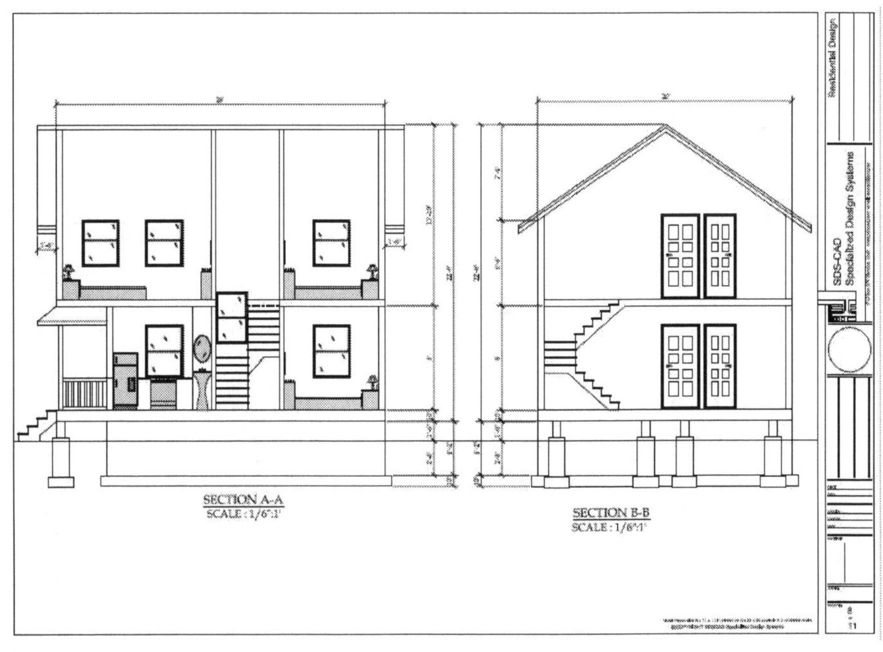
SECTION A-A
SCALE : 1/8"=1'
SECTION B-B
SCALE : 1/8"=1'
Residential Design
SDS-CAD
Specialized Design Systems

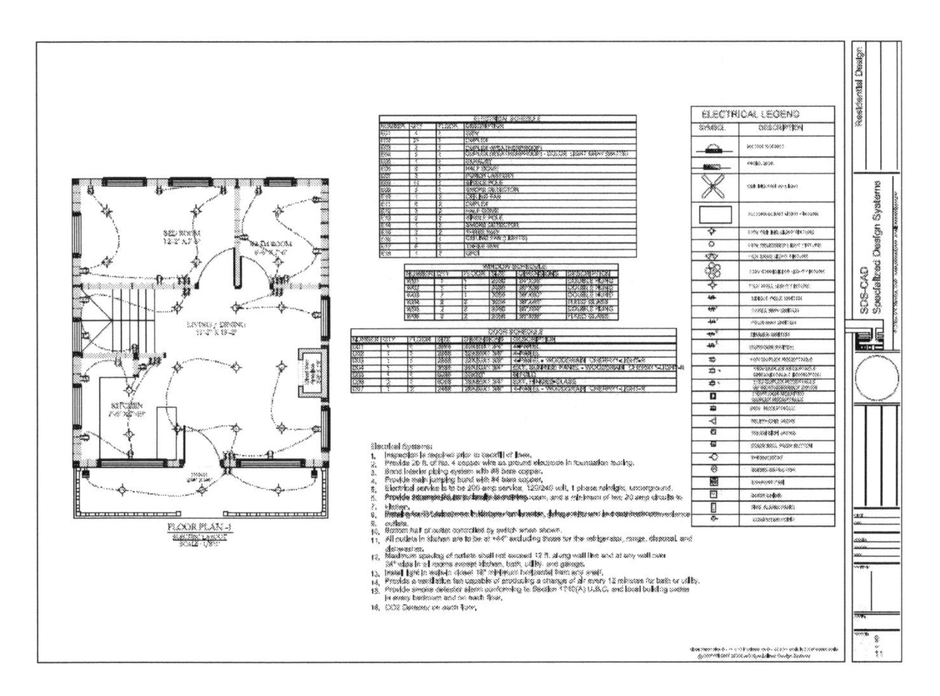

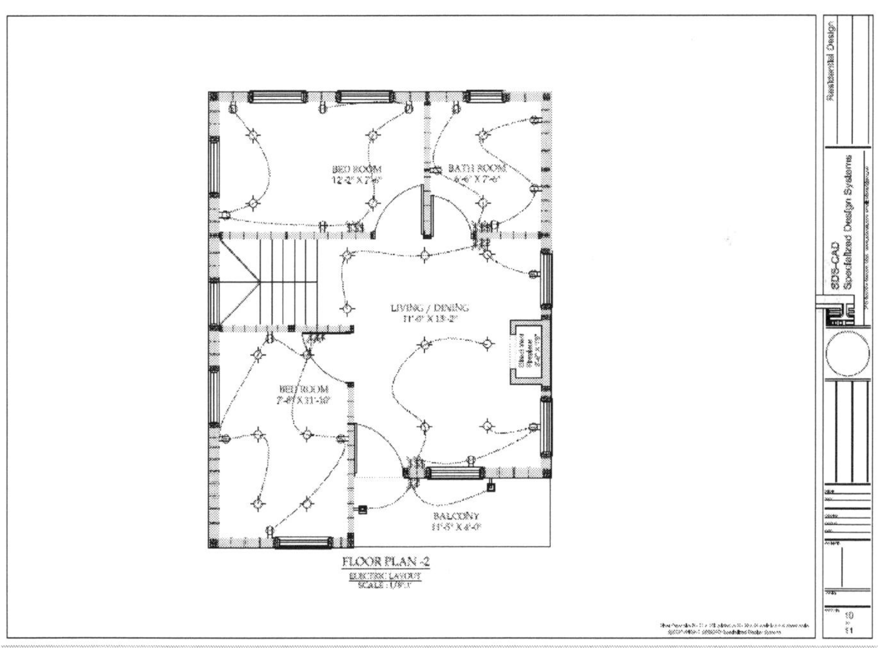
BED ROOM
12'-2" X 7'-6"
BATH ROOM
LIVING / DINING
BED ROOM
BALCONY
FLOOR PLAN -2
ELECTRIC LAYOUT
Residential Design
SDS-CAD
Specialized Design Systems

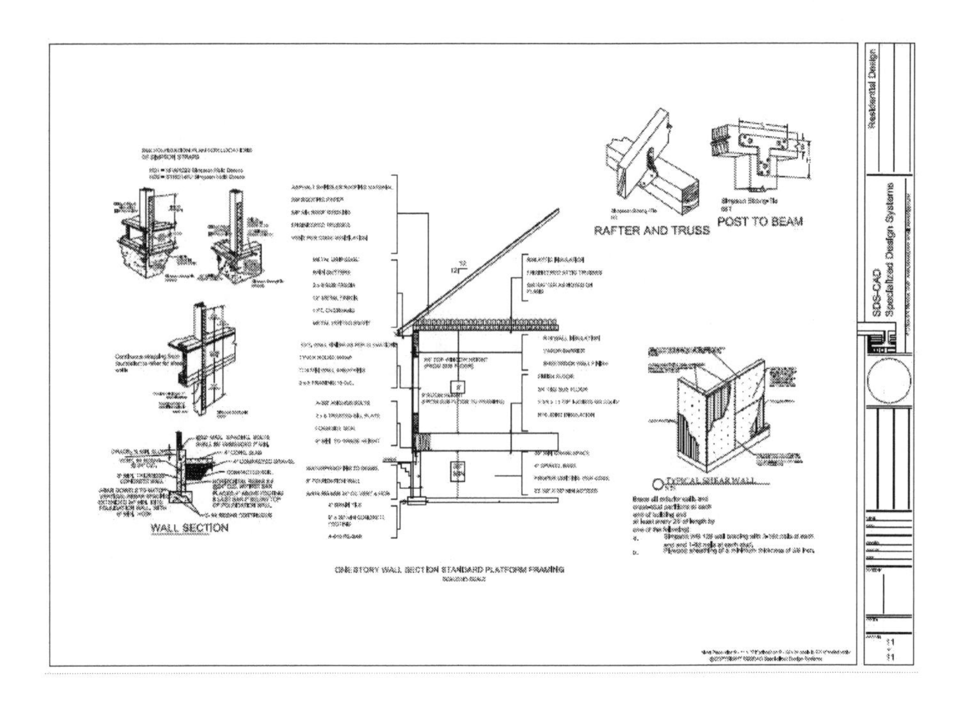
RAFTER AND TRUSS
POST TO BEAM
WALL SECTION
TYPICAL SHEAR WALL
ONE STORY WALL SECTION STANDARD PLATFORM FRAMING
Residential Design
SDS-CAD
Specialized Design Systems

House #10: 621 ft^2

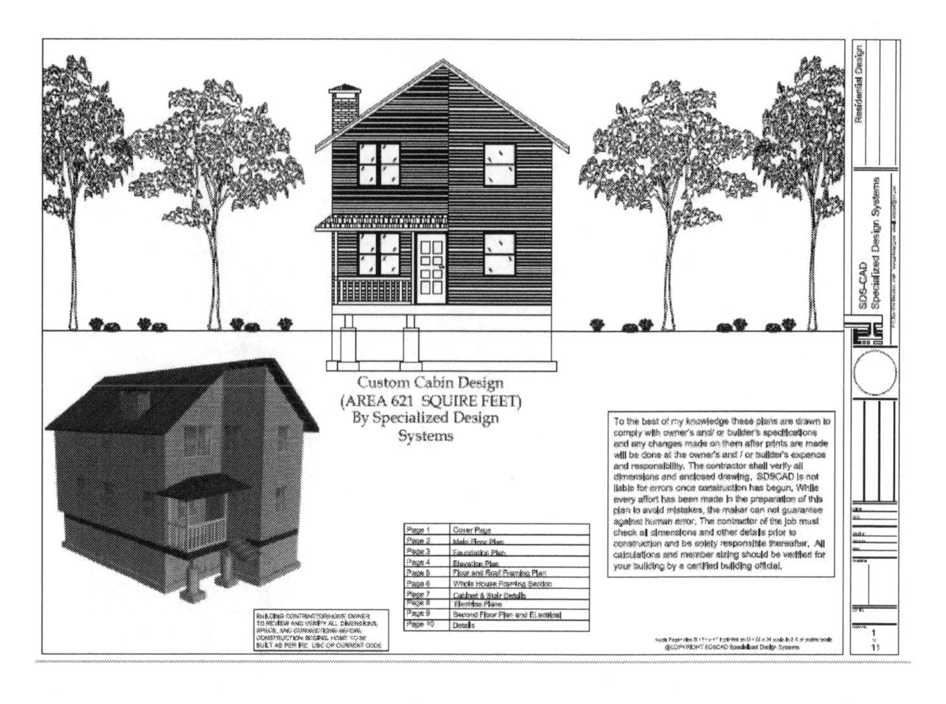
Custom Cabin Design
(AREA 621 SQUIRE FEET)
By Specialized Design
Systems
To the best of my knowledge these plans are drawn to comply with owner's and/ or builder's specifications and any changes made on them after prints are made will be done at the owner's and / or builder's expence and responsibility. The contractor shall verify all dimensions and enclosed drawing. SDSCAD is not liable for errors once construction has begun. While every effort has been made in the preparation of this plan to avoid mistakes, the maker can not guarantee against human error. The contractor of the job must check all dimensions and other details prior to construction and be solely responsible thereafter. All calculations and member sizing should be verified for your building by a certified building official.
Page 1 Cover Page
Page 2 Main Floor Plan
Page 3 Foundation Plan
Page 4 Elevation Plan
Page 5 Floor and Roof Framing Plan
Page 6 Whole House Framing Section
Page 7 Cabinet & Stair Details
Page 8 Electrical Plans
Page 9 Second Floor Plan and Electrical
Page 10 Details
Residential Design
SDS-CAD
Specialized Design Systems

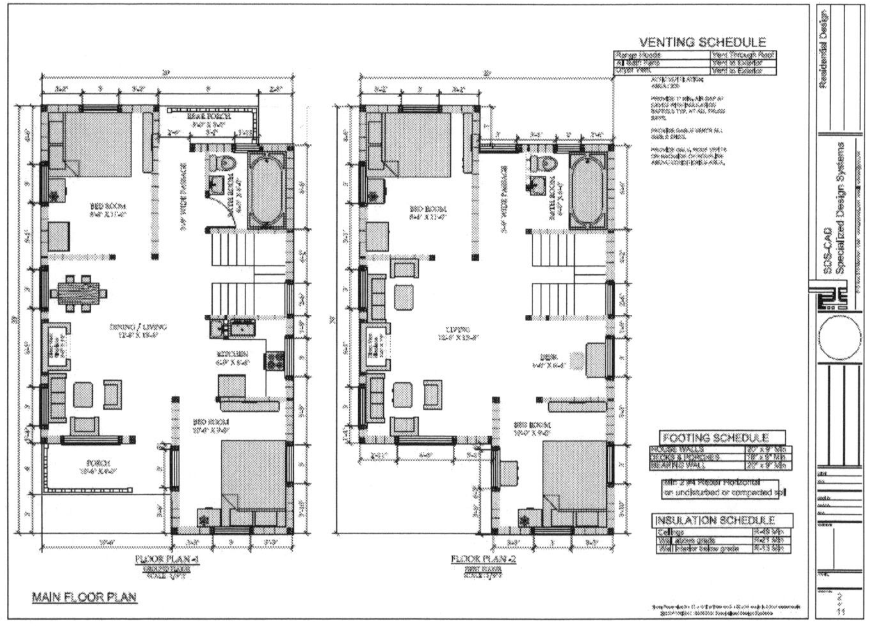
VENTING SCHEDULE
Range Hoods | Vent Through Roof
All Bath Fans | Vent to Exterior
Dryer Vent | Vent to Exterior
REAR PORCH
BED ROOM
BATH ROOM
3'-6" WIDE PASSAGE
DINING / LIVING
KITCHEN
PORCH
LIVING
DESK
FOOTING SCHEDULE
HOUSE WALLS
DECKS & PORCHES
BEARING WALL
Min. 2 #4 Rebar Horizontal
on undisturbed or compacted soil
INSULATION SCHEDULE
Ceilings
Wall above grade
Wall interior below grade
FLOOR PLAN -1
FLOOR PLAN -2
MAIN FLOOR PLAN
Residential Design
SDS-CAD
Specialized Design Systems

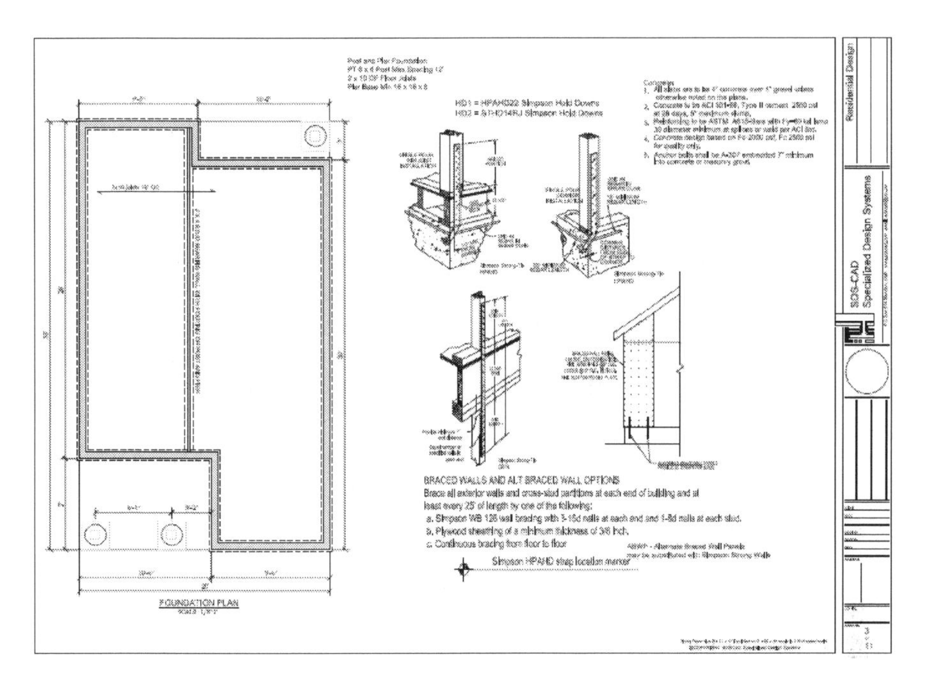
Post and Pier Foundation
HD1 = HPAHD22 Simpson Hold Downs
HD2 = STHD14RJ Simpson Hold Downs
BRACED WALLS AND ALT BRACED WALL OPTIONS
Brace all exterior walls and cross-stud partitions at each end of building and at least every 25' of length by one of the following:
a. Simpson WB 126 wall bracing with 3-16d nails at each end and 1-8d nails at each stud.
b. Plywood sheathing of a minimum thickness of 3/8 inch.
c. Continuous bracing from floor to floor
Simpson HPAHD strap location marker
FOUNDATION PLAN
Residential Design
SDS-CAD
Specialized Design Systems

METAL ROOF
GRADE
FRONT VIEW 1-1
LEFT VIEW 3-3
Wood or Vinyl
Siding over
structural panel
Residential Design
SDS-CAD
Specialized Design Systems

Residential Design
SDS-CAD
Specialized Design Systems
METAL ROOF
GRADE
BACK VIEW 2-2
RIGHT VIEW 4-4
Wood or Vinyl
Siding over
structural panel

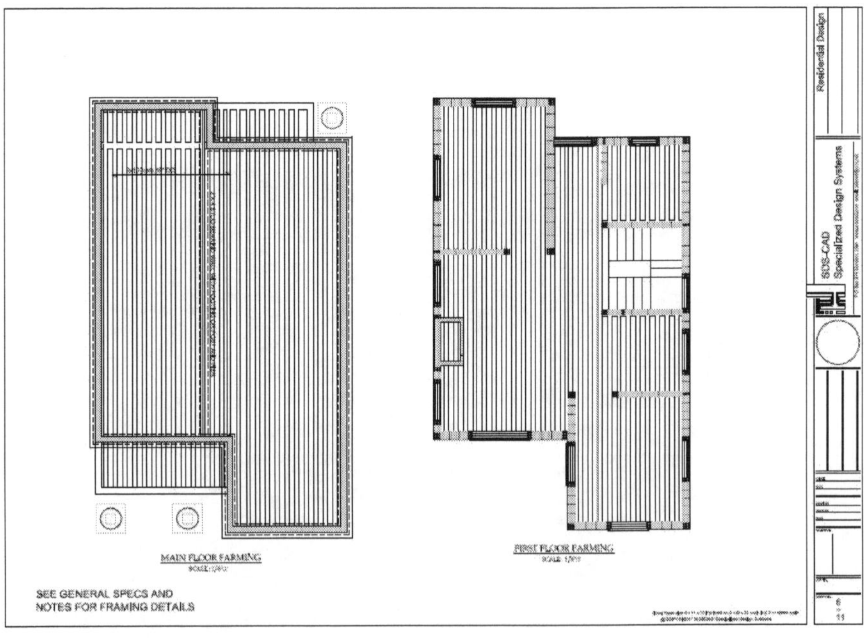
Residential Design
SDS-CAD
Specialized Design Systems
MAIN FLOOR FARMING
FIRST FLOOR FARMING
SEE GENERAL SPECS AND
NOTES FOR FRAMING DETAILS

Roof Framing:

2. For soffit size see details.
3. For spans and dimensions refer to floor plans.

SEE GENERAL SPECS AND
NOTES FOR FRAMING DETAILS

12/8 PITCH
1 1/2 x 11 1/2 I-Joists or DF 2 x 12
3 x 11 1/2 LVL Ridge Beam
ROOF FARMING

Residential Design

SDS-CAD
Specialized Design Systems

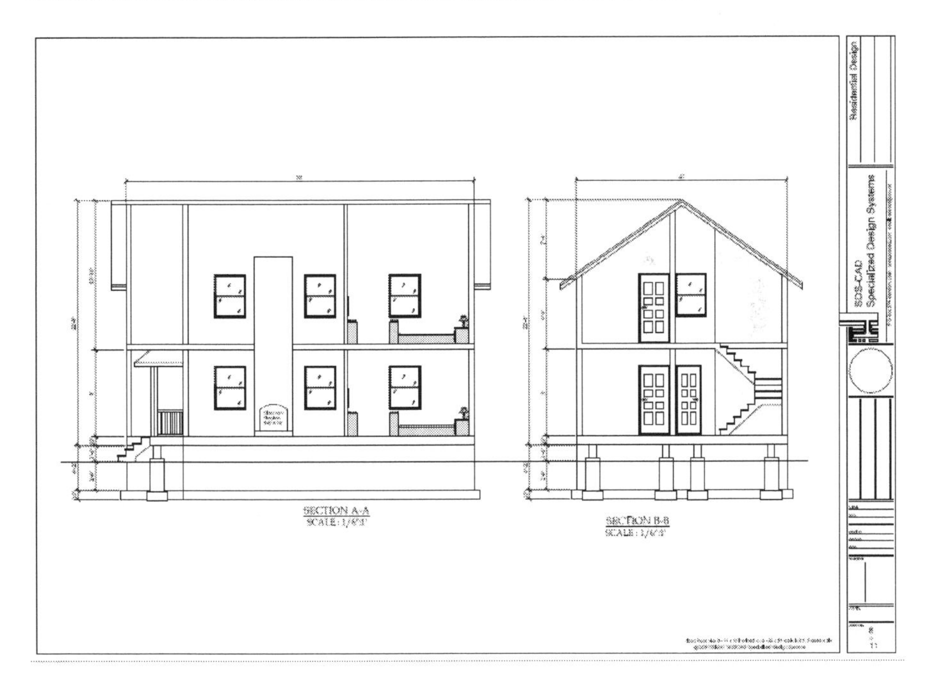

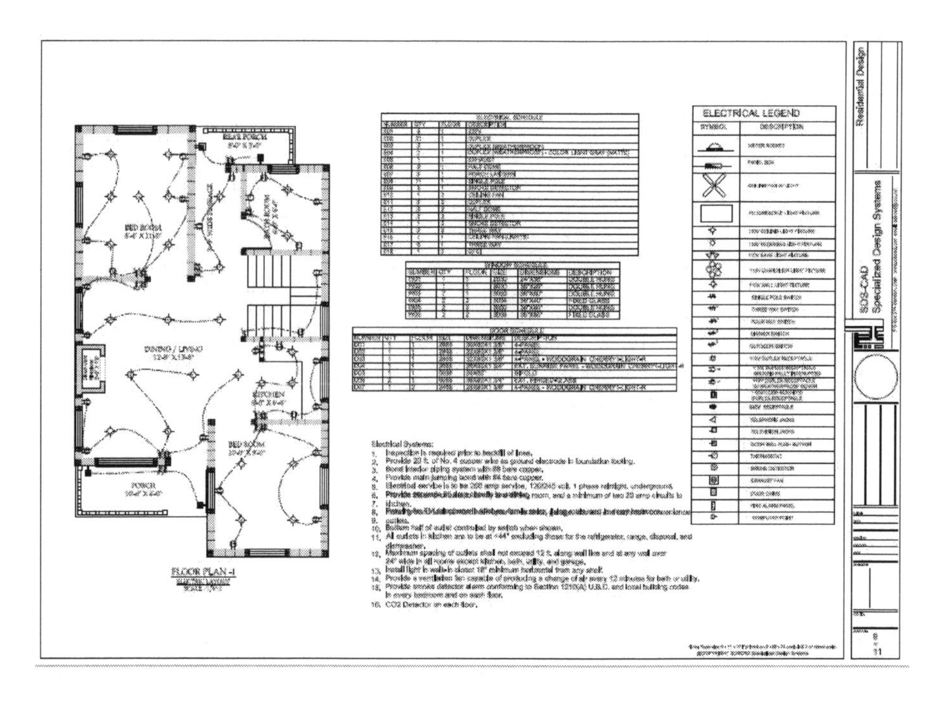
ELECTRICAL LEGEND
DINING / LIVING
BED ROOM
FLOOR PLAN -1
Residential Design
SDS-CAD
Specialized Design Systems

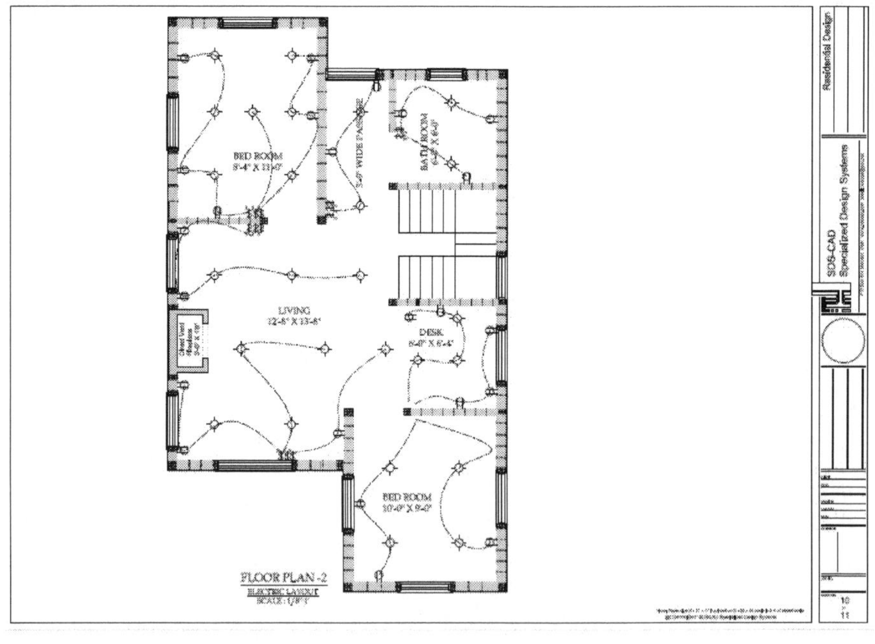
BED ROOM
8'-4" X 11'-0"
BATH ROOM
LIVING
12'-8" X 13'-8"
DESK
BED ROOM
10'-0" X 9'-0"
FLOOR PLAN -2
Residential Design
SDS-CAD
Specialized Design Systems

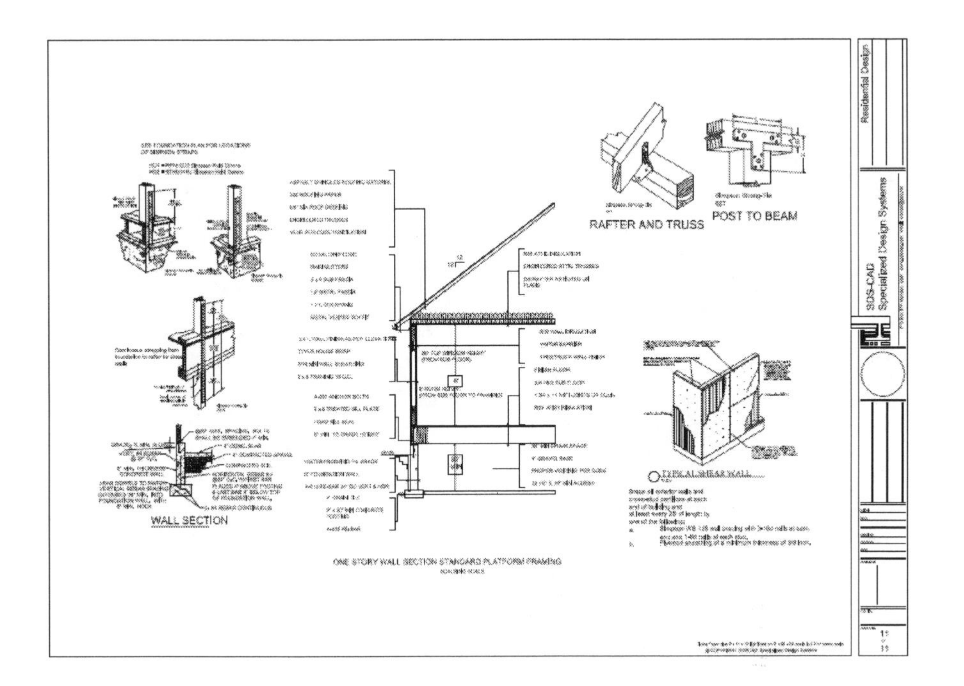
RAFTER AND TRUSS
POST TO BEAM
WALL SECTION
ONE STORY WALL SECTION STANDARD PLATFORM FRAMING
TYPICAL SHEAR WALL
Residential Design
SDS-CAD
Specialized Design Systems

Portable House #1

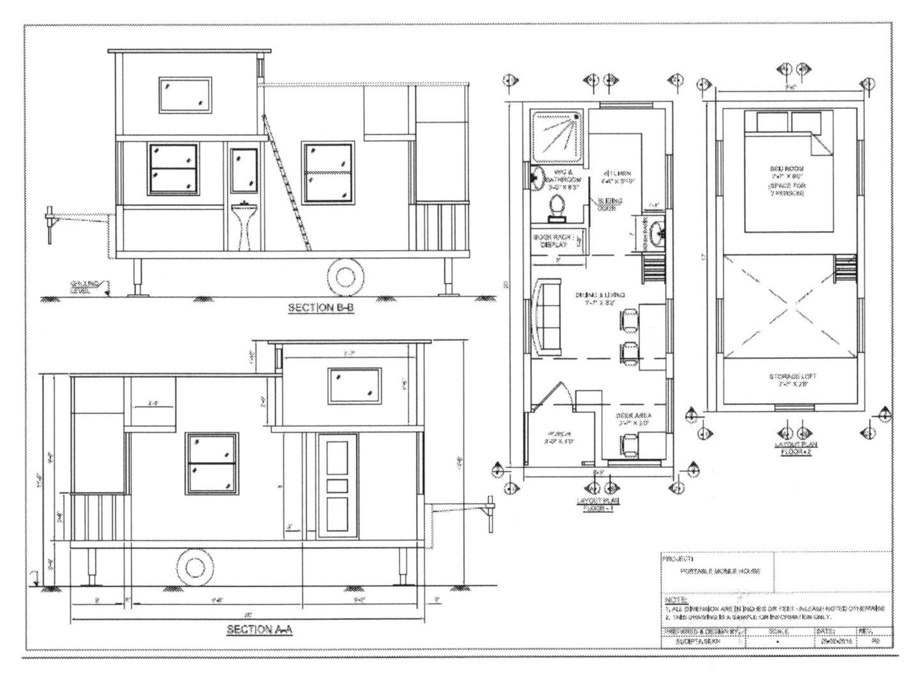
SECTION B-B
SECTION A-A

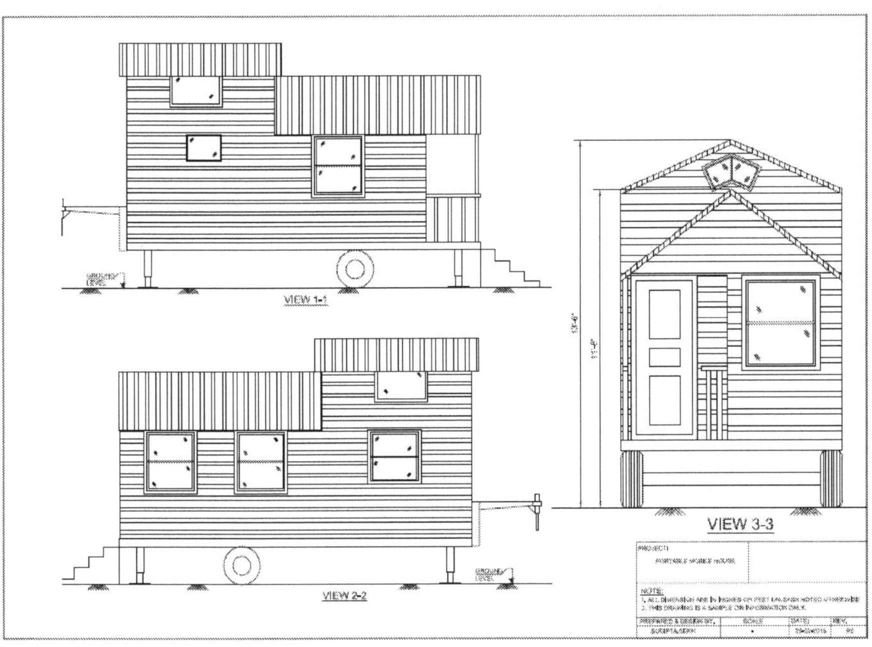
VIEW 1-1
VIEW 3-3
VIEW 2-2

Portable House #2

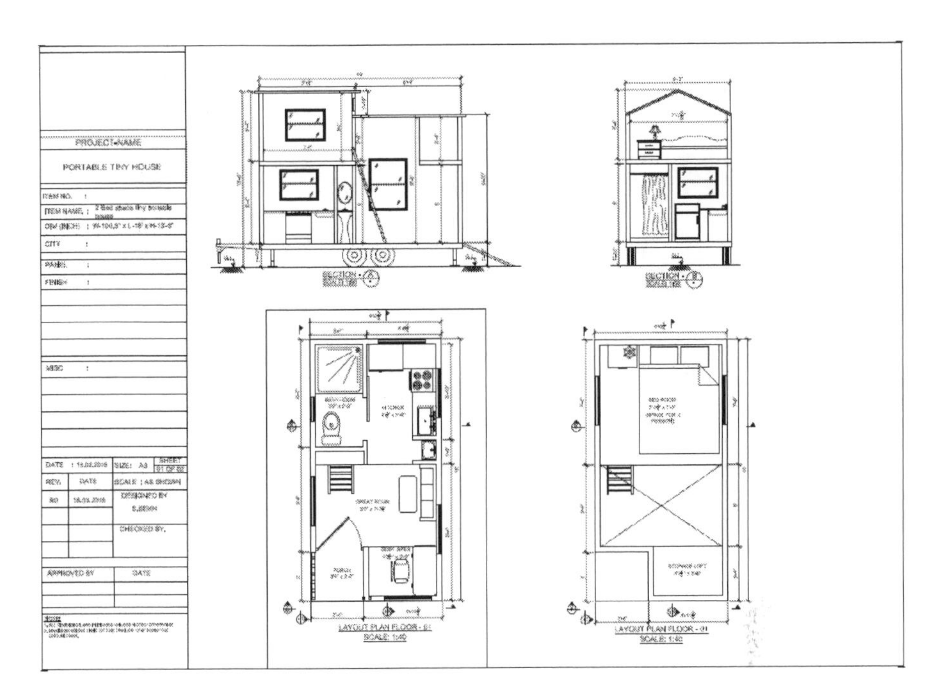
PROJECT-NAME
PORTABLE TINY HOUSE
LAYOUT PLAN FLOOR - 01
SCALE: 1:40

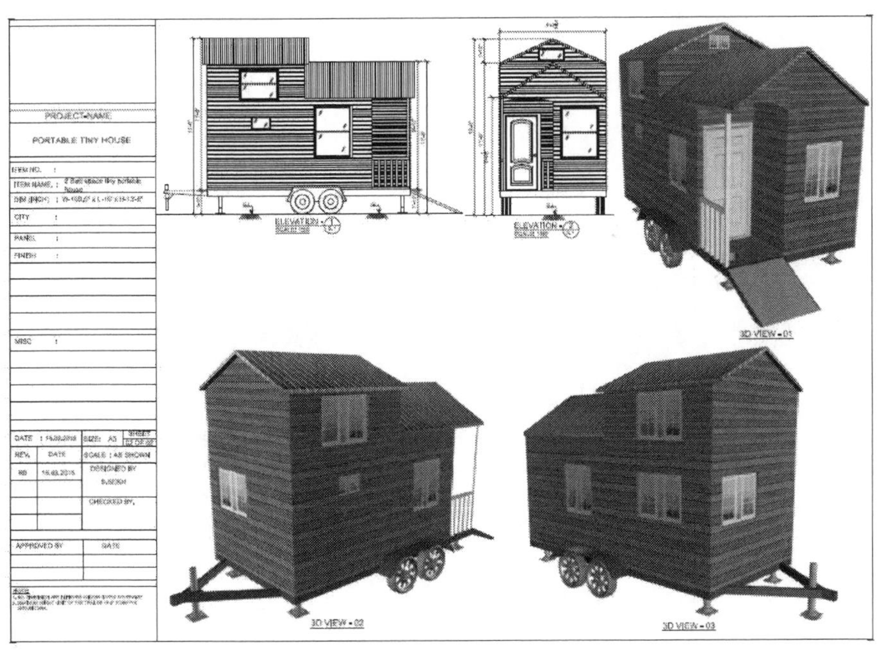
PROJECT-NAME
PORTABLE TINY HOUSE
ELEVATION
3D VIEW - 01
3D VIEW - 02
3D VIEW - 03

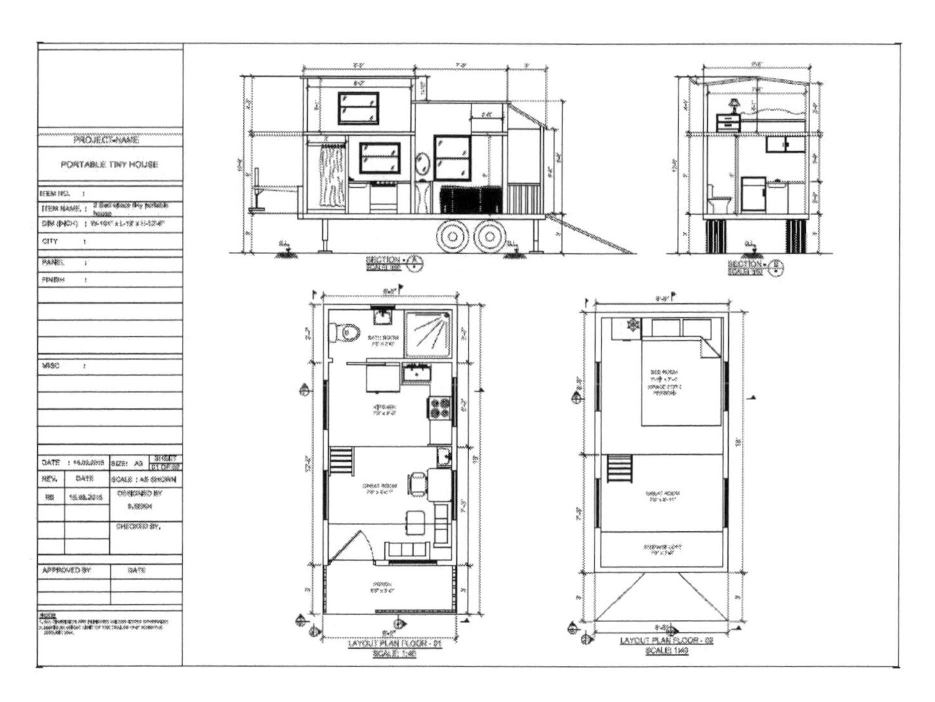
PROJECT NAME
PORTABLE TINY HOUSE
ITEM NO. :
CITY :
PANEL :
FINISH :
MISC :
CHECKED BY.
APPROVED BY
DATE
SECTION
LAYOUT PLAN FLOOR - 01
LAYOUT PLAN FLOOR - 02

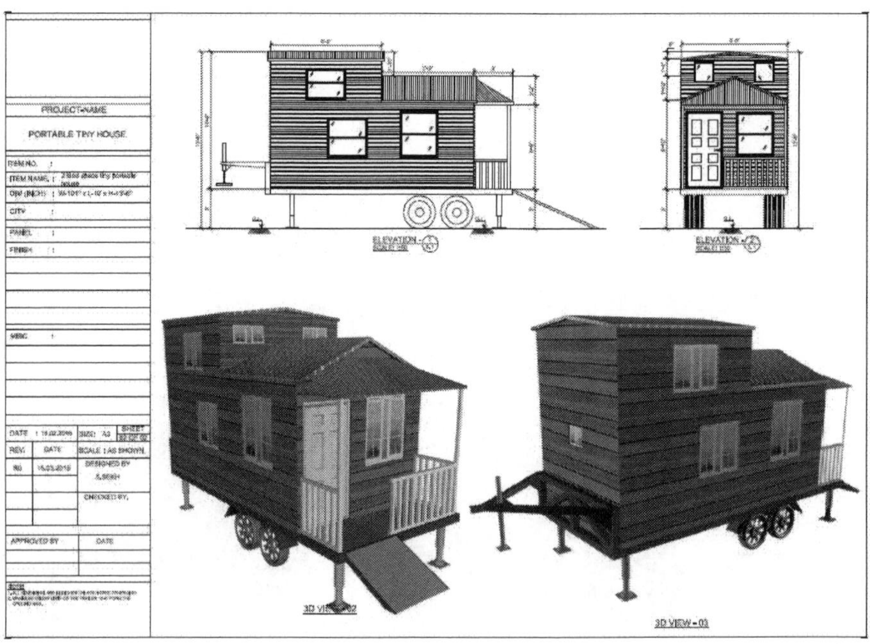
PROJECT NAME
PORTABLE TINY HOUSE
ITEM NO. :
CITY :
PANEL :
FINISH :
MISC :
CHECKED BY.
APPROVED BY
DATE
ELEVATION
3D VIEW - 03

Resource Guide

Company Name Website

Pinterest - Portable Small Home Ideas

www.pinterest.com/campsitelm/portable-small-home-ideas/

Tumbleweed - Tiny House Company

www.tumbleweedhouses.com

Gizmag - Tiny Leaf House can accommodate a family of four

www.gizmag.com/leaf-house-tiny-portable-home/22861/

RPC - Builder of Tiny Homes and Cabins

www.richsportablecabins.com

Designboom - Tiny Houses

www.designboom.com/contemporary/tiny_houses2.html

Houzz - Small Portable Houses

www.houzz.com/small-portable-houses

Busyboo - Small Portable House-To-Go

www.busyboo.com/2008/08/05/small-house-togo/

TrendHunter - 41 Transportable Homes

www.trendhunter.com/slideshow/transportable-homes

TinyHouseTalk - Portable and Foldable Tiny House

www.tinyhousetalk.com/portable-and-foldable-tiny-house/

American Tiny House - Tiny House Plans on Wheels

www.americantinyhouse.com/tiny-house-plans-on-wheels/

Tiny Smart House

www.tinysmarthouse.com

TinyHousePins - Homebox 1 Portable Three Story Tiny House

www.tinyhousepins.com/homebox-1-portable-two-story-tiny-house/

Choice Home Warranty - 21 Unique Portable Homes

www.choicehomewarranty.com/blog/21-unique-portable-homes/

Tiny House Giant Journey

www.tinyhousegiantjourney.com

Tiny House Design

www.tinyhousedesign.com

Tiny House Blog

www.tinyhouseblog.com

Austin Tiny House

www.austintinyhouse.com

Brevard Tiny House

www.brevardtinyhouse.com

Tiny Green Cabins

www.tinygreencabins.com

Tiny Home Builders

www.tinyhomebuilders.com

Tiny Idahomes LLC

www.tinyidahomes.com

Trekker Trailers

www.trekkertrailers.com

Turtle Island Tiny Homes

www.turtleislandtinyhomes.ca

Small House Society

www.smallhousesociety.net

PAD Portland Alternative Dwellings

www.padtinyhouses.com

Resilience - Building a Tiny House

www.resilience.org/resource-detail/2544932-building-a-tiny-house

tinyHouseBuild.com

www.tinyhousebuild.com

Tiny Home Resource

www.tiny-home-resource.com

East Coast Tiny Homes

www.eastcoasttinyhomes.com

Sustainable, Non-Toxic, Ecowise Tiny Homes

www.ecowisetinyhomes.com

Humble-Homes - Tiny House Plans by Humble Homes

www.humble-homes.com/tiny-house-plans/

Little House on the Trailer

www.littlehouseonthetrailer.com

Rocky Mountain Tiny Houses

www.rockymountaintinyhouses.com

Shelter Wise

www.shelterwise.com

Tennessee Tiny Homes

www.tinyhappyhomes.com

Tiny House Company

www.tinyhouseco.com

Vermont Tiny Houses

www.vermonttinyhouses.com

Wishbone Tiny Homes

www.wishbonetinyhomes.com

Check out some of the other JD-Biz Publishing books

Gardening Series on Amazon

Download Free Books!

http://MendonCottageBooks.com

Health Learning Series

GRANDMA'S
NATURAL REMEDIES
AND ANCIENT HERBAL
BEAUTY RECIPES
Volume 1
HEALTH LEARNING SERIES
DILEEP J SINGH AND J DAVIDSON
GRANDMA'S
NATURAL REMEDIES
AND ANCIENT HERBAL
BEAUTY RECIPES
HEALTH LEARNING SERIES
DILEEP J SINGH AND J DAVIDSON
GRANDMA'S
NATURAL REMEDIES
AND ANCIENT RECIPES
GRANDMA'S CURE FOR OBESITY
GRANDMA'S CURE FOR THE COMMON COLD
Volume 3
HEALTH LEARNING SERIES
DILEEP J SINGH AND J DAVIDSON
GRANDMA'S
NATURAL REMEDIES AND
ANCIENT HERBAL RECIPES
Volume 4
HEALTH LEARNING SERIES
DILEEP J SINGH AND J DAVIDSON
GRANDMA'S
HERBAL LORE
ANCIENT HERBAL RECIPES AND REMEDIES
Volume 5
HEALTH LEARNING SERIES
DILEEP J SINGH AND J DAVIDSON
GRANDMA'S
ANCIENT
BEAUTY REMEDIES
FROM HER KITCHEN
Volume 6
HEALTH LEARNING SERIES
DILEEP J SINGH AND J DAVIDSON
GRANDMA'S
EASY TO USE TIPS
IN THE KITCHEN AND OUTDOORS
Volume 7
HEALTH LEARNING SERIES
DILEEP J SINGH AND J DAVIDSON
GRANDMA'S
HOUSEHOLD HINTS AND RECIPES
USING TIME TESTED
ECONOMICAL TIPS IN YOUR HOME
75 Tips
Remove Stains
Peel Tomatoes
Perfect Pies
HEALTH LEARNING SERIES
DILEEP J SINGH AND J DAVIDSON
GRANDMA'S
NATURAL REMEDIES
AND ANCIENT RECIPES
ALL 5 BOOKS IN 1
HEALTH LEARNING SERIES
DILEEP J SINGH AND J DAVIDSON

Country Life Books

Health Learning Series

Amazing Animal Book Series

Learn To Draw Series

How to Build and Plan Books

Entrepreneur Book Series

Our books are available at

1. Amazon.com
2. Barnes and Noble
3. Itunes
4. Kobo
5. Smashwords
6. Google Play Books

Publisher

JD-Biz Corp

P O Box 374

Mendon, Utah 84325

http://www.jd-biz.com/

Made in the USA
San Bernardino, CA
03 August 2016